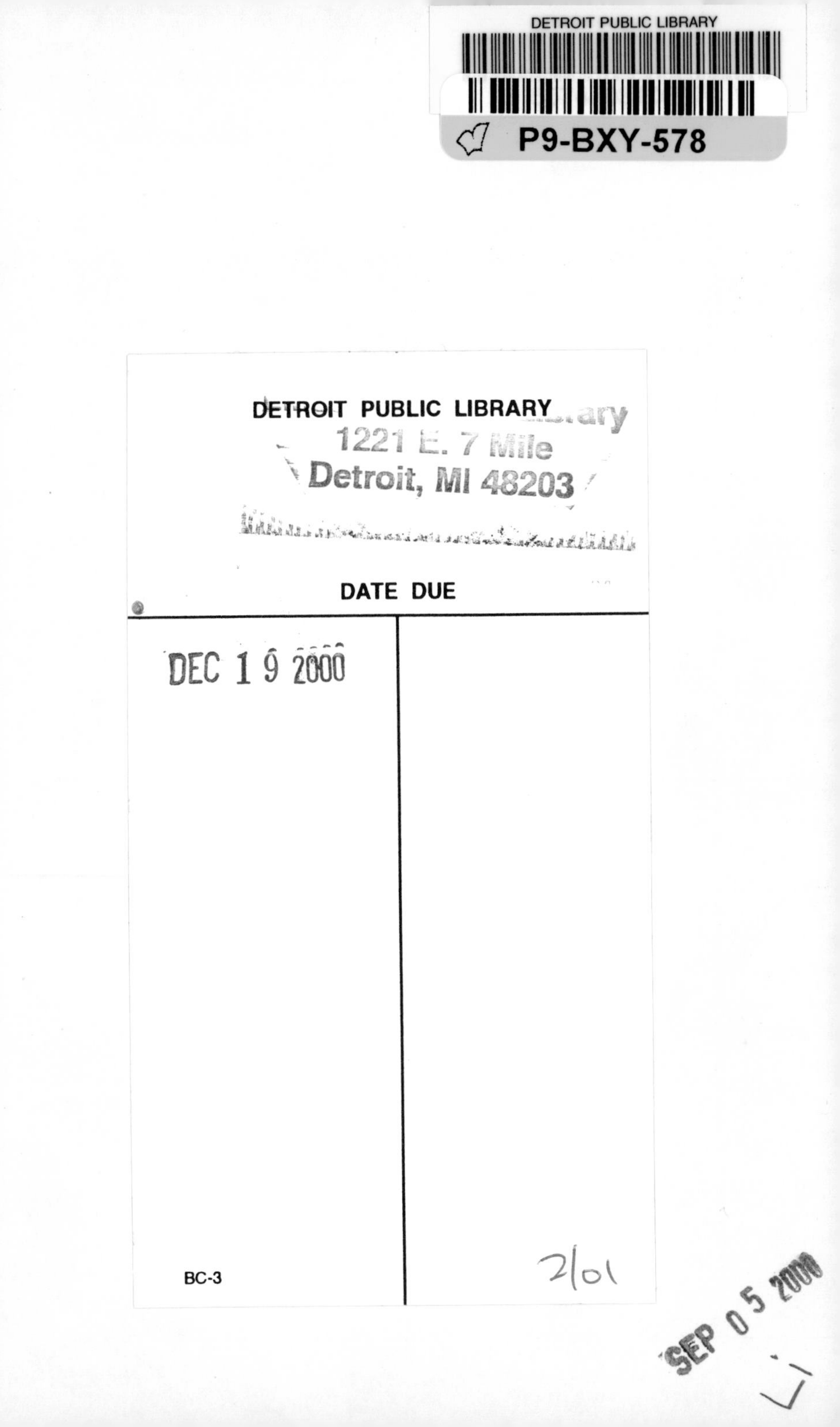

cool chemistry

great experiments with simple stuff

Steven W. Moje

Sterling Publishing Co., Inc.
New York

Library of Congress Cataloging-in-Publication Data Available

Moje, Steven W.

10 9 8 7 6 5 4 3 2 1

Published by Sterling Publishing Company, Inc.
387 Park Avenue South, New York, N.Y. 10016

Distributed in Canada by Sterling Publishing
c/o Canadian Manda Group, One Atlantic Avenue, Suite 105
Toronto, Ontario, Canada M6K 3E7
Distributed in Great Britain and Europe by Chris Lloyd
463 Ashley Road, Parkstone, Poole, Dorset, BH14 0AX, England
Distributed in Australia by Capricorn Link (Australia) Pty Ltd.,
P.O. Box 6651, Baulkham Hills, Business Centre, NSW 2153, Australia
Manufactured in the United States of America

Sterling ISBN 0-8069-6349-2

Contents

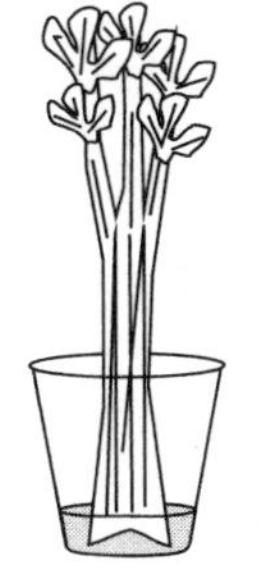
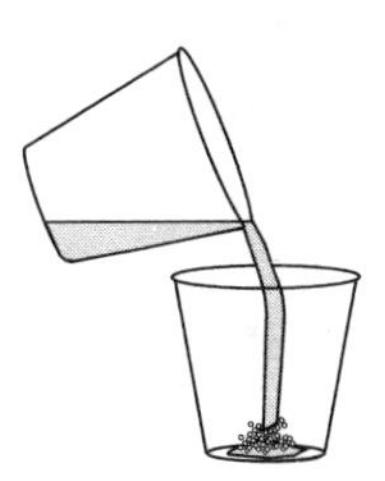

Preface

The physical world is made up of matter. Matter is something that takes up space, which we can sense by sight, hearing, smell, taste, or touch. The chemical elements are the building blocks of matter. Chemistry is the physical science that describes the properties of substances and how substances change when they react with each other. A physical change (such as water freezing into ice) is reversible. Ice can easily be changed back into water by applying a small amount of heat. On the other hand, a chemical change (such as a nail's rusting) is not reversible. Rust (iron oxide) cannot be changed back into an iron nail, no matter how hard you try.

Whatever we can see, touch, taste, smell, or hear involves chemistry. Our physical surroundings, our clothing, our homes, the food we eat, the water we drink, and our physical bodies — all involve chemistry. We sometimes hear on television and read in newspapers that chemicals are bad. It is true that certain chemical substances can hurt us,but many are very important, even essential for life itself, such as water to drink and oxygen in the air we breathe. *Cool Chemistry* is a hands-on beginner's chemistry book for children of ages 8 to 12, or anyone who enjoys science fun. It includes fun, easy, safe chemistry experiments, done using ordinary things found in the home. None of the experiments involves flames or excessively hazardous substances. Normal adult supervision is recommended to ensure that the experiments in this book are done properly. You will learn as you do the experiments in this book that chemistry is not only very interesting, but also a lot of fun.

Definitions

Matter is something that takes up space, which we can sense by sight, hearing, smell, taste, or touch.

States of Matter

Gases, such as air, have no definite volume (size) or shape.

Liquids, such as water, have a definite volume, but no definite shape.

Solids, such as iron, have both a definite volume and definite shape.

Changes in States of Matter: Physical Changes

Melting is a change from a solid to a liquid that happens when a solid is heated.

Subliming is a change from a solid to a gas that happens when a solid is heated.

Freezing is a change from a liquid to a solid that happens when a liquid is cooled.

Boiling is a change from a liquid to a gas that happens when a liquid is heated.

Evaporation is a change from a liquid to a gas that happens when a liquid is standing.

Condensation is a change from a gas to a liquid that happens when a gas is cooled.

Distillation means heating an impure liquid to boiling,

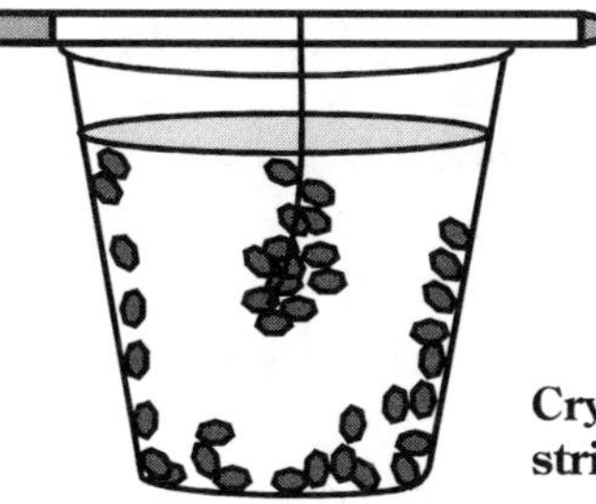

Crystals forming on a string (see page 36)

then cooling the resulting vapors (gas) and collecting the pure liquid.

Filtration is the separation of a solid from a liquid by passing a suspension of both through a paper or cloth.

A *mixture* is a combination of pure substances that keep their identity and can be separated by physical means (such as distillation, evaporation, or filtration). The amount of each substance in a mixture can vary. Examples of mixtures are air, salt water, and sand in water. Air contains the gases nitrogen, oxygen, argon, and carbon dioxide. These air gases can be separated by distillation. Salt water is a solution of salt in water, and is separable by evaporation or by distillation. Sand in water can be separated by filtration.

An *emulsion* is a liquid that contains finely dispersed, very tiny particles.

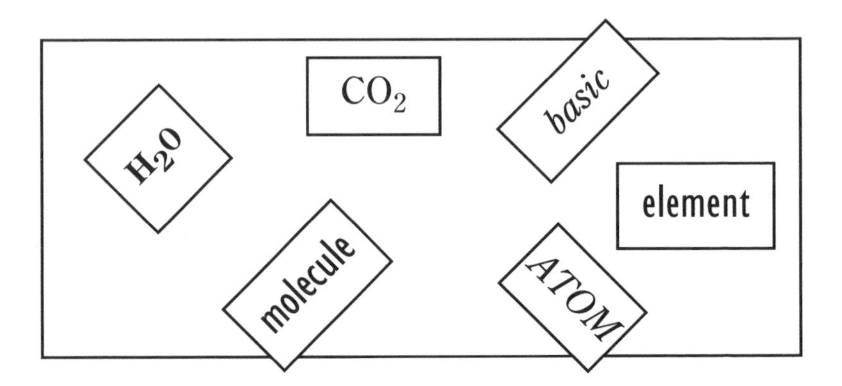

Chemical Symbols, Formulas, and Terms

An *element* is a substance that cannot be chemically divided into any simpler substance. There are about 100 elements. Examples of elements are carbon, hydrogen, helium, oxygen, iron, and lead.

An *atom* is the smallest particle of an element that still has the properties of an element. Atoms are made up of

even smaller parts, which include negatively charged electrons whirling about the positively charged nucleus or center of the atom. Sometimes atoms give up or gain electrons when they join with other atoms.
A *molecule* is a unit of two or more atoms that are chemically joined together. A molecule could have only one kind of atom in it, like an oxygen molecule, or two or more different kinds of atoms, like a water molecule. A water molecule consists of two atoms of hydrogen and one atom of oxygen.
An *ion* is an atom, group of atoms, or a molecule that has a positive or negative electrical charge. The charge is indicated by a + or - sign.
A *compound* is a substance made of two or more elements in a fixed ratio. The elements of a compound cannot be separated by physical means. A compound has different properties from the elements of which it is made. Examples of chemical compounds are water (composed of hydrogen and oxygen) and sugar (composed of carbon, hydrogen, and oxygen).
Unstable. Some compounds are unstable, which means they are very likely to change into something else.

Scientists have developed a shorthand name for each element, and shorthand ways of noting how many atoms of each are in a molecule of a substance.
Each element has one or two letters representing it. Here are some you may see in this book: C (carbon), O (oxygen), H (hydrogen).
The number of atoms of each element is indicated in a chemical substance's formula. The formula tells you how many of each kind of atom join together to make the smallest unit of the chemical substance. For example, CO_2, a molecule of carbon dioxide, has one carbon atom and two oxygen atoms.

Below are definitions of some other terms you will see in this book:

An *acid* is a substance that ionizes (gives up electrons) in water to form H^+ (hydrogen) ions. Some acids are weak. They are only partially ionized in water. Some are strong; they are completely or almost completely ionized in water. Vinegar contains a weak acid, acetic acid.

A *base* is a substance that produces hydroxide (OH^-) ions in water. When we say a substance is *basic,* it has a special meaning in chemistry. It means that something behaves like a base.

A *salt* is an ionic compound whose ions are neither H^+ nor OH^-. Salts are formed when acids and bases react together to neutralize each other.

An *indicator* is a substance that shows by a change in its color whether it is in an acidic or basic substance.

Equipment List

Here's a list of the things you need to do the experiments in this book. Each project has its particular list also.

Aluminum foil • Bag, dark-colored
Baking tray • Balloons, large and small round
Ballpoint pen • Bottle, 1- or 2-liter plastic soda bottle
Bowl • Cloth, scrap
Coat hanger
Coins (U.S. pennies dated before and after 1982, or other copper coins)
Containers for food
Cups (paper, plastic, Styrofoam, 3 to 9 oz)

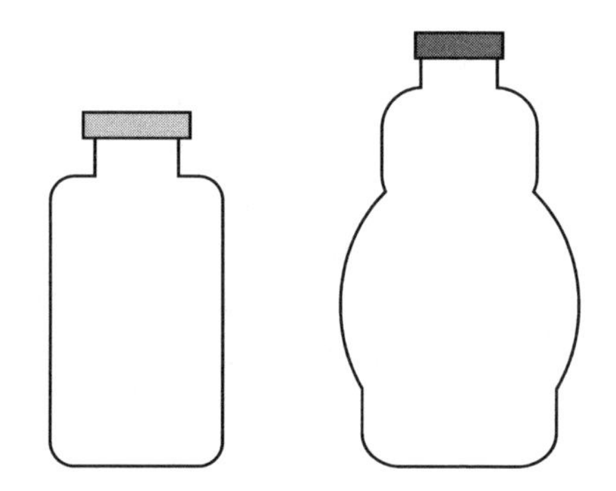

Coffee filters (paper, cone style)
Coffee stirring sticks (hollow)
Dishpan, plastic • Duct tape • Flashlight
Flour • Fork •Funnel
Glasses (paper, plastic, or glass drinking glasses)
Grater (cheese or other food) • Ice cube tray
Index cards, paper
Iron-containing objects (nails, steel wool, steel-wool cleaning pads)
Jars with lids
Knife (plastic or metal table or kitchen knife)
Markers (permanent and water-soluble, black and various other colors)
Measuring cup • Milk container (1-gallon plastic)
Modeling clay • Paintbrush (small)
Paper clips, medium (#1 size) • Paper towels
Pencil • Pipe cleaners
Plastic sandwich bag, with resealable zipper closing
Plastic wrap
Plates (ceramic, plastic, Styrofoam dinner plates)
Ice cream bar sticks (craft sticks) • Rocks (tiny)
Rubber bands (medium to large)

Supply List

Ammonia (household)
Antacid tablets, fizzing (like Alka-Seltzer®)
Baby powder (cornstarch-based)
Baking powder • Baking soda (sodium bicarbonate)
Bleach, solid (fabric bleach) • Body powder
Candies, coated with colored sugar shells
Carbonated beverages (soda pop): cola, non-cola, and carbonated (seltzer) water
Cookies • Cream • Cream of tartar
Detergent for automatic dishwasher

Detergent, laundry (with and without bleaching agent)
Detergent, liquid dishwashing • Earth (soil)
Eggs • Epsom salt (magnesium sulfate)
Flowers • Food coloring, artificial (red, blue, green, yellow) • Fruits (see project to learn which ones you need) • Gelatin (unflavored and flavored)
Glue (school glue) • Honey • Ice •
Iodine tincture
Fruit juices (see projects for details)
Ice cream bars, colored
Juice drink (like Kool-Aid®)
Leaves from tree
Legumes, dried (beans, lentils, peas)
Milk (4%, 2%, 1%, skim milk) • Milk of magnesia
Nuts and peanuts (see projects for details)
Pepper (black, white, or red) • Popcorn (unpopped)
Rice, dry (white, short and long grain, instant, brown)
Rubbing alcohol (isopropyl alcohol)
Salt (kosher, regular, and iodized) • Sand
Seeds (bird and sunflower) • Soap, bar
Sour salt (citric acid) • Spaghetti (dry)
Spices: extracts (mint, vanilla, orange, lemon, and banana); powdered, colored (cumin, curry, and turmeric)
Sugar (granulated, powdered, brown, cube)
Vegetables (see projects for details)
Vegetable oils (see projects for details)
Vegetable shortening, lard or meat fat
Vinegar labeled clear, white, or distilled
Water (tap water and distilled or deionized)
Yeast, active dry

Safety Precautions

Never drink or eat the substances used for any of your experiments or allow them to splash into your eyes, mouth, or ears. Although most of the substances used in this book are completely safe, several (household ammonia, iodine tincture, laxative tablets, and rubbing alcohol) do require certain precautions:

*It's a good idea when working with irritating substances such as household ammonia and iodine tincture to wear safety glasses or goggles and a plastic apron to protect against splashes.

*Household ammonia has a nasty, pungent odor and is dangerous if it gets into your eyes. Do not leave the bottle open, and do not smell the fumes. For maximum safety, wear safety goggles and a plastic apron. Do not use any stronger kind of ammonia for these experiments.

*Iodine tincture, a wound medicine, is a strong irritant. Do not touch it or allow it to get on your skin.

*Rubbing alcohol is flammable and toxic. Do not allow it to get near a flame, and do not taste it.

*Be sure to keep all chemicals and equipment out of the reach of small children. If you leave an experiment somewhere, label it so no one will hurt himself with it or throw it out.

Experiments with Physical Properties

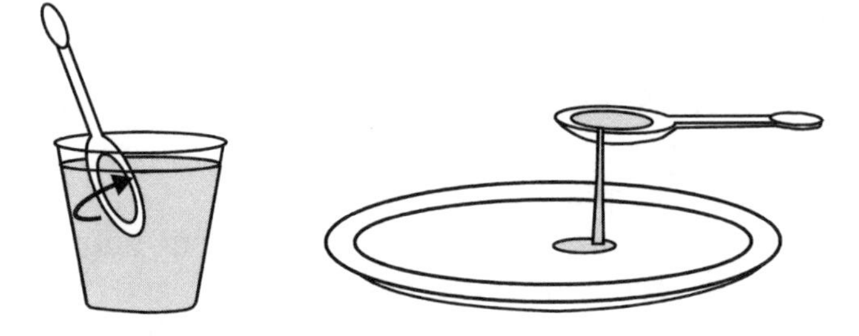

Physical Change: Salt Dissolved in Water

You will need: water, salt, cup, teaspoon, dinner plate, ice cube tray, refrigerator

What to do: Add a teaspoon of salt to a cup of water. Stir until the salt is dissolved. Place a spoonful of this liquid on a dinner plate. Allow the plate to sit for several days, until all of the water has evaporated and a white substance remains on the plate. What remains is salt.

How it works: When salt is added to water, a solution of salt water forms. The dissolved salt can be easily turned back into its original (solid) state by allowing salt water to stand for several days. The water, which evaporates into the air as a gas, will return to its original liquid state later when it falls as rain.

A physical change occurs when a substance changes its state or form, but does not become a new chemical substance. Salt dissolved in water has gone through a physical change. Physical changes are reversible, unlike chemical changes, which are not normally reversible.

More science fun: Pour water into an ice cube tray and place the tray in the freezer compartment of your refrigerator. In just a few hours, the water will freeze into ice cubes, which are solids. An ice cube can easily change back to its original liquid state if you put it on your kitchen table and allow it to melt. This easy reversibility (water changing to ice, then back to water again) proves that freezing is a physical and not a chemical change.

Chemical Change: Vinegar and Baking Soda

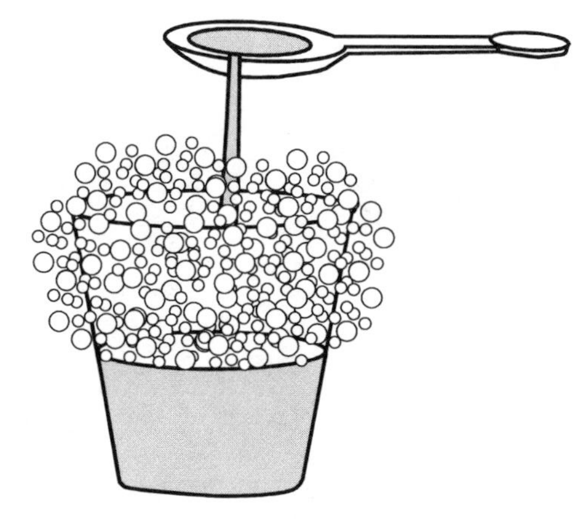

You will need: cup or drinking glass, teaspoon, vinegar, baking soda

What to do: Half-fill a cup or drinking glass with vinegar. Add a teaspoon of baking soda to the glass of vinegar. The liquid will foam over the top of the cup!

How it works: Vinegar contains an acid (acetic acid) and baking soda is a base. When these two chemicals contact each other, they form a salt called sodium acetate and a weak acid called carbonic acid (H_2CO_3). Carbonic acid is *unstable** and falls apart into water (H_2O) and carbon dioxide (CO_2). Carbon dioxide, a gas, causes the liquid in the glass to foam. Since vinegar reacting with baking soda is a *chemical* change, it is not reversible (unlike the example on page 14: water changing to ice, and then ice changing back to water again).

**Unstable:* Tending to change.

More chemistry fun: Add different amounts of baking soda to vinegar. How many spoonfuls of baking soda give the most foam? Try this experiment with warm vinegar. Does warm vinegar give more foam with baking soda than room-temperature vinegar does? Add a drop of dish detergent to the vinegar. Does this give even more foam?

sand + water

sand + salt

salt + water

Mixtures and Solutions: Sand, Salt, and Water

You will need: 4 jars, water, sand, salt, coffee filter

What to do: Shake sand and water together in a jar for a few seconds. Stop. The sand will drop to the bottom of the jar. In a second jar, shake salt and sand together. After you stop shaking this jar, you will be able to see separate grains of salt and sand. In a third jar, mix a small amount of salt with water and shake it. The salt will dissolve and seem to disappear.

How it works: Sand + water and salt + sand are mixtures. In a mixture, the particles do not dissolve into each other. They keep their own identities. Salt dissolving in water to give salt water is an example of a solution. A solution is a special kind of mixture in which one substance dissolves in the other.

More science fun: Add water to the mixture of salt and sand (second jar). The salt will dissolve, but the sand, which is not soluble in water, remains on the bottom of the jar. The sand can be separated from this jar of salt water by filtering the contents through a coffee filter. When the salt water that was passed through the filter is allowed to evaporate, you can recover the salt. This is an excellent and easy way to separate salt from sand. Try it with beach sand. Compare beach sand from the ocean (where it was wet with salt water) and from lakes (wet with freshwater). Which type of sand gives more residue after being mixed with water, filtered and evaporated? Why?

Weighing Scale

You will need: wire coat hanger, 2 paper cups, modeling clay or aluminum foil, string, many coins or paper clips, ballpoint pen or toothpick

What to do: You can make your very own scale to weigh things! Punch three holes in the rims of two paper cups. Use a ballpoint pen or toothpick to most easily and safely make these holes. The holes should be equal distances from each other around the rim of the cup. Put a knot at one end of each of six pieces of string that are about 8 inches (20 cm) long, and push the unknotted end of each string through one of the holes in the cup, working

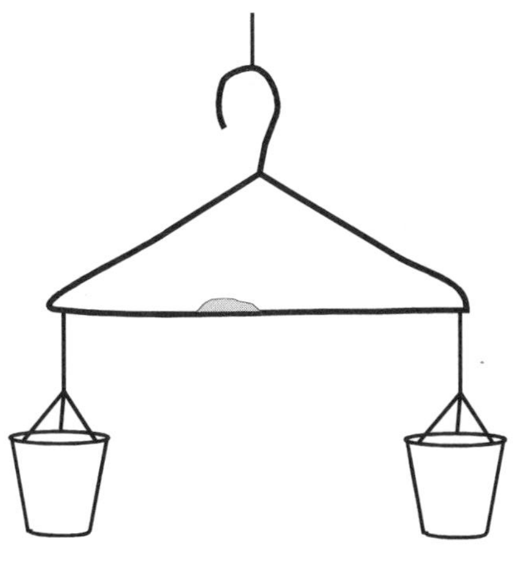

from the inside of the cup outwards. About 3 inches (7 cm) above each cup, tie together the three ends of the string from that cup, as shown in the drawing. Tie two pieces of string about 10 inches (25 cm) long to the triangular corners of the coat hanger; then tie each corner string to the knot joining the three strings over a cup. Tie another piece of string around the coat hanger hook. Hang the coat hanger from the edge of a table by tying the coat hook string around something heavy, such as a board, and resting it on the edge of the table. Place a piece of clay or a lump of aluminum foil near the center of the coat hanger and move the clay or foil until the coat hanger is level.

To weigh something, place a small object in one of the cups. That cup will go down, and the other (empty) cup will go up. Place small objects such as coins or paper clips into the empty cup until this second cup sinks to the same height as the cup with the object in it.

How it works: Gravity pulls down on the objects in the cups with equal force. When the coins or paper clips in the second cup exactly balance the object in the first cup, the two sides have the same weight. Instead of traditional weight units (ounces or pounds in the U.S.A., or grams or kilograms in most of the rest of the world) you can think in terms of coin units. For example, if a small rock weighs 10 coins, a bigger rock twice as heavy as the first rock would weigh 20 coins.

More science fun: Weigh a variety of small objects with your scale, such as small pieces of fruit (cherries and raisins), pencils, gumballs and other candies, erasers, and ice cubes.

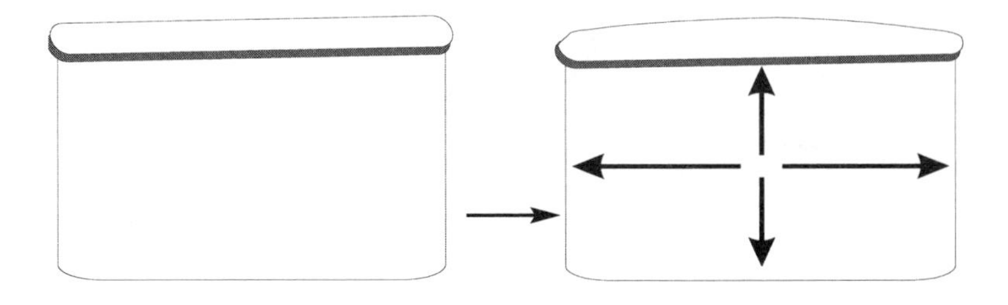

Expandable Air

You will need: food container with tight-fitting lid, warm to hot water.

What to do: Pour a small amount of hot water in a food container. Quickly put the lid on the container. Notice that the lid is flat. Now quickly shake the container. The lid will bulge up!

How it works: When you shake the food container, the hot water warms the air inside the container. Warm air occupies more space than cold air. When (relatively) cool air is sealed into a container and then suddenly warmed, the air expands. The bulging of the lid shows that the air pressure has increased.

More science fun: Try different sizes of containers. How much more does hot water raise the lid compared with warm water? What happens when you nearly fill up the container with a lot of hot water and then shake it? Does the lid bulge as much as when the container had very little hot water in it?

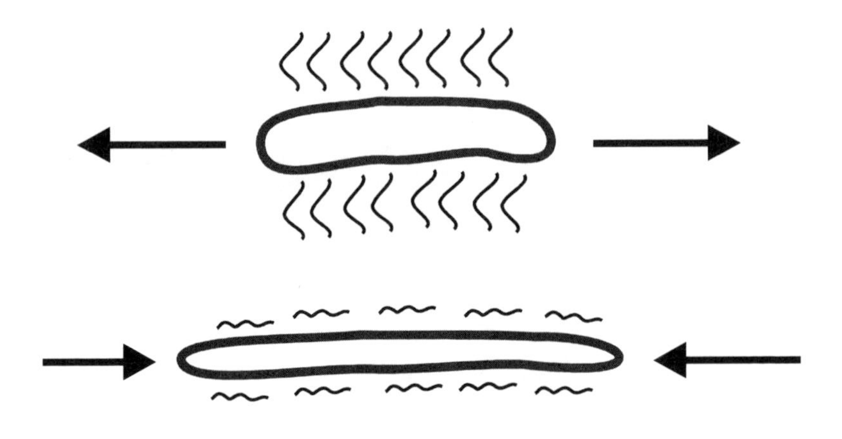

Hot and Cold Rubber Band

You will need: rubber bands (medium to large size)
What to do: Place the index finger each hand inside the loop of a medium to large size rubber band. Pull your hands apart slightly, just enough to keep the rubber band from slipping off your hand. Hold the rubber band up against your lips. Suddenly pull your hands apart, stretching the rubber band. The rubber band will get warm! Hold the rubber band stretched against your lips for 10 seconds, allowing it to cool off. Now, still holding the stretched rubber band against your lips, suddenly push your hands toward each other. The rubber band will become quite cold!
How it works: A rubber band becomes hot when stretched and cold when released from its stretched state because of a physical law called *entropy*. Entropy is a fancy way of saying how random or scattered something is. When you are playing in your bedroom and have toys all over the place, your bedroom has high amount of entropy. When you have cleaned up your room and put all your toys away, your room has low

entropy. Dominos, when arranged neatly in vertical rows, have low entropy. When you push one domino into the other dominos, they all fall over; the fallen dominos have high entropy. When you stretch the rubber band across your lips and pull it tight, the coils in the rubber become stretched and forced into a line (like the vertical dominos). When a chemical substance such as rubber is stretched and forced to become less random (low entropy), it releases heat. When you suddenly release the stretched rubber band, its molecules return from their ordered (low-entropy) stretched state to their preferred (high-entropy) random coiled state. In order for the rubber band to become more random, heat energy must be absorbed from another object or substance.

Human lips are very sensitive to small changes in hot and cold and can easily detect a rubber band's release of, and absorption of, heat. Your lips (in addition to the air) transmit heat to and from the rubber band.

More science fun: Vary the experiment by stretching and releasing the rubber band at differing speeds and also by using different-sized rubber bands. You will be amazed at how much a difference in heat gained and lost can be detected, depending upon the how fast the rubber band is stretched and released, and also depending on the size (width, length, and thickness) of the rubber band.

Blow a Balloon into a Bottle

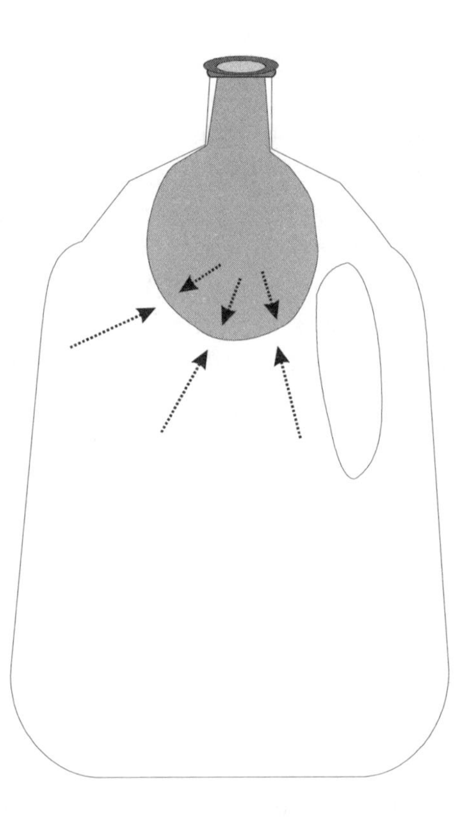

You will need: Two 1-gallon (or 2-liter) plastic milk jugs, large round balloon

What to do: Push a large round balloon into the neck of a plastic milk jug. Stretch the neck of the balloon over the mouth of the milk jug. Try to inflate (blow up) the balloon in the milk jug. You will only be able to blow it up slightly (if at all).

How it works: Air pressure trapped in the milk jug pushes back against the balloon as you blow into the balloon. This prevents your inflating the balloon.

More science fun: Cut a large hole in the side or bottom of the milk jug with scissors. You should now be able to easily inflate the balloon inside the jug. Why? In a second milk jug, poke a very small hole by puncturing the jug with the tip of a ballpoint pen. Is it more difficult to inflate the balloon in this jug than in the first jug? How big a hole is needed to easily get the balloon to inflate?

Straw Sprayer

You will need: soda straw, drinking glass, water, mint or vanilla extracts, food coloring (blue and yellow)

What to do: Cut a soda straw in half. Stick one of the straw halves (#1) vertically into a glassful of water. Place the other straw piece (#2) against the first straw piece at a right angle. Blow through it. The harder you blow through straw #2, the higher the water will rise in straw #1. When water reaches the top of the straw #1, a fine mist of water will spray all over the place!

How it works: Moving air has less pressure than still air. When you blow through the horizontal straw, across the top of the vertical straw, the air pressure on the surface of the water in the glass (at the bottom of straw #1) is greater than the air pressure at the top of the straw. The greater the pressure difference, the higher the water is able to rise in the straw. For best spraying results, make sure that the straws are slightly touching, and that the lower edge of straw #2 is even with the top of the vertical straw. If the straws are not lined up properly, the air pressure difference will not be enough to pull the water up high enough to be sprayed.

More science fun: Cut different lengths of straws. Which length gives the best spraying? Make a perfume sprayer by adding a few drops of mint or vanilla extract to the water. For a magic color-changing water sprayer, add some blue food coloring to the water. Spray the blue water onto some yellow food coloring placed in a dish. The yellow color will turn green!

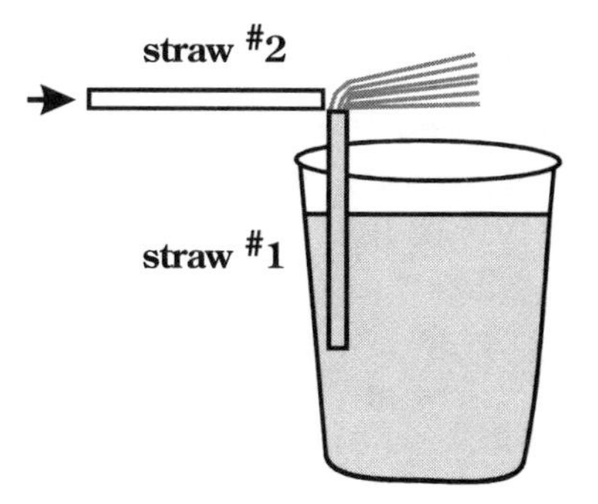

Spoon Glued to Water

You will need: water faucet, a few spoons

What to do: Hold the convex (rounded outside) surface of a spoon against the stream of water from a water faucet. The water will grab onto the spoon, pulling it against the water stream, almost as if the water were gluing the spoon!

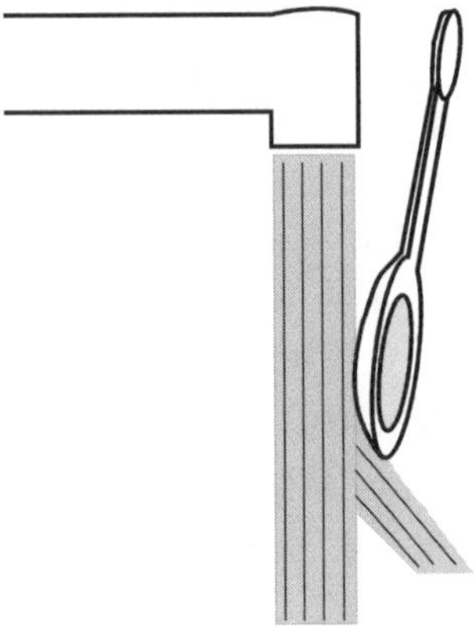

How it works: The spoon is pulled into the stream of water because a moving liquid (water) has lower pressure than the air that surrounds the spoon.

More science fun: Try different rates of water flow. How fast does the water flow have to be before you feel the spoon being drawn into the water? Try using different kinds of spoons, such as tablespoons, teaspoons, measuring spoons, and spatulas. Does the size or shape of the spoon affect how strongly the spoon is attracted to the stream of water?

Styrofoam Water Buddies

You will need: small (one-eighth-inch) pieces of Styrofoam (made by breaking the top edge from a Styrofoam cup and crumbling the rim into small pieces), tiny scraps of paper or ground pepper, 2 whole Styrofoam cups, toothpick, water, dishwashing liquid detergent

What to do: Half-fill a Styrofoam cup with water.

Sprinkle the cup pieces on the water surface of the cup. Most of the pieces will travel to the sides of the cup. Carefully and completely fill a second cup with water until the water level is slightly *above* the rim of the cup. Sprinkle small pieces of Styrofoam on the water surface of this cup. Most of the pieces will move to the *middle* of the cup!

How it works: Styrofoam is attracted to paper and plastic but is repelled by (pushes away from) water. In the half-full cup, the Styrofoam pieces move to the sides and away from the center, in order to avoid water as much as possible. In the over-full cup (where the Styrofoam pieces cannot touch the edge of the cup at all), the pieces move to the center, as close to each other as possible, still trying to avoid water.

More science fun: Use a toothpick to move the Styrofoam pieces away from the sides of the half-full cup. The pieces will move right back to the edge of the cup! What happens when you try to push the pieces away from the center of the over-full cup? Add a few drops of dish detergent to both cups. Does this affect how the floating pieces go to the edges of the half-full cup or to the center of the over-full cup? Repeat the experiment with tiny scraps of paper or ground pepper. Does paper or pepper behave as dramatically as do pieces of Styrofoam?

Water Bulge

You will need: 3-oz paper cup, water, many paper clips or other small heavy objects, such as tiny rocks

What to do: Carefully and completely fill a 3-oz cup to the brim with water. After it is full, slowly and gently add paper clips or other small heavy objects, such as tiny rocks, one at a time. Look at the cup from the side; the water will actually rise *above* the top of the cup!

How it works: The surface tension (the tendency of a liquid to hold itself together) of water is strong enough to overcome the force of gravity, which tries to pull the water over the edge of the cup. This surface tension creates a thin water "skin," which you can see bulging above the top of the cup when paper clips or other small heavy objects are added. After the water gets so high above the rim that it cannot fight gravity any longer, it spills over the top of the cup.

More science fun: How many paper clips can you add before the water spills over the edge? If you are using tiny rocks instead, which work better: clean rocks or dirty rocks?

Densities of Liquids

You will need: water, vegetable oils (corn, olive, sesame, peanut, and mixtures), rubbing alcohol, salt water, sugar water, honey, dish detergent, juices, milk (skim, 1% fat, 2% fat, whole (4% fat), cream, clear drinking glass, 3-oz paper cups, ruler, marker, weighing scale (see page 17)

What to do: Fill a clear drinking glass half-full with water. Add some drops of vegetable oil to it. The oil will float on the surface. Now fill a 3-oz paper cup half-full (about 1 inch from the top) with vegetable oil. Add a few drops of water to the oil. The water will go to the bottom of the cup.

How it works: A given volume of oil weighs less than the same volume of water. Another way of saying this is that the density of oil is less than that of water. Oil wants to float on top of water, no matter how much or little there is of either oil or water.

More science fun: You can easily compare the densities of any two liquids by weighing equal volumes. Use the weighing scale you made earlier in this book (page 17) to compare relative weights. Measure several 3-oz cups with a ruler and mark a line 1 inch from the bottom of the cup (about one-half the height of the cup) with a marker. Pour water up to the 1-inch mark in one of the cups. The water is the standard you will use to compare the relative weights of other liquids. Some liquids to compare with water are: vegetable oils (corn, olive, sesame, peanut, and mixtures), rubbing alcohol, salt water, sugar water, honey, dish detergent, juices, milk (skim, 1%, 2%, 4%), and cream. Take turns pouring each of

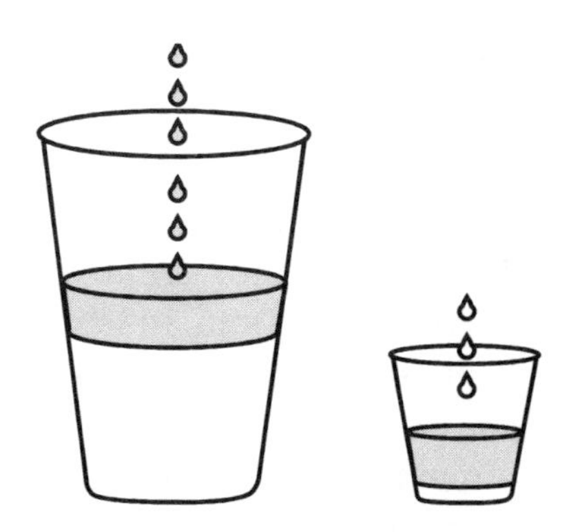

these liquids up to the 1-inch mark in another cup. Test them on your weighing scale with an equal volume of water on one side. Which liquids weigh more than the same volume of water? Which weigh the same as water? Which weigh less than the same volume of water?

Oil Globules and Emulsions

You will need: clear drinking glasses, straws or coffee stirrer sticks, different kinds of vegetable oils (corn, olive, sesame, peanut, and mixtures), water, rubbing alcohol, salt water, sugar water, dish detergent, laundry detergent, fruit and vegetable juices, milk (skim, 1%, 2%, and 4% fat), and cream

What to do: Add several drops of vegetable oil to a glass of water. They will form droplets (globules) that float around on the surface of the water. Watch the globules of oil for up to a minute, noticing if any globules dissolve. Repeat the experiment with glasses of liquids other than water, such as rubbing alcohol, salt water, sugar water, dish detergent, fruit and vegetable juices, milk (skim, 1%, 2%, whole {4%}), and cream. In some liquids, oil droplets will break up and dissolve. In other liquids (such as water) oil droplets will float together for a long time.

How it works: The more that a liquid is unlike oil, the longer that the oil will stick together and not dissolve. The more that the liquid is like oil, the more quickly the oil will dissolve. Water (denser than oil) doesn't tend to remain in contact with oil. On the other hand, detergent loves oil. Even though the density of detergent is different from oil's, because detergent's chemical structure encourages it to be attracted to oil, the two mix together well. And because detergent also likes water, it pulls oil and water together to form an emulsion. An emulsion is a liquid that contains very tiny, finely dispersed particles.

More chemistry fun: Does any liquid keep oil droplets together (undissolved) longer than water does? Which liquid dissolves oil droplets most quickly? Try using different kinds of vegetable oils (corn, olive, sesame, peanut, and mixtures). Which kinds of oil make the longest-lasting globules in water? Try different types of detergents to see which gives the best emulsion. Which ratios of oil/detergent/water give the best (longest-lasting) emulsions?

Floating and Sinking Water Balloons

You will need: 2 large plastic dishpans, 2 small water balloons, water (cold, room temperature, and warm)

What to do: Fill the first dishpan with warm water and the other pan with cold water. Fill two small water balloons with room-temperature water; make sure that no air remains in either of the balloons. Place one of the balloons in the first (warm water) dishpan. The balloon will sink. Place the other balloon in the second (cold water) dishpan. The balloon will float.

How it works: Warm water is less dense than cold water. The colder the water in the balloon, compared to the pan water, the more easily the balloon sinks in water that is warmer (and therefore less dense) than it is. The warmer the water in the balloon, the more easily the balloon floats in water that is cooler (and therefore denser) than it is.

More science fun: After balloon #2 has been in the cold

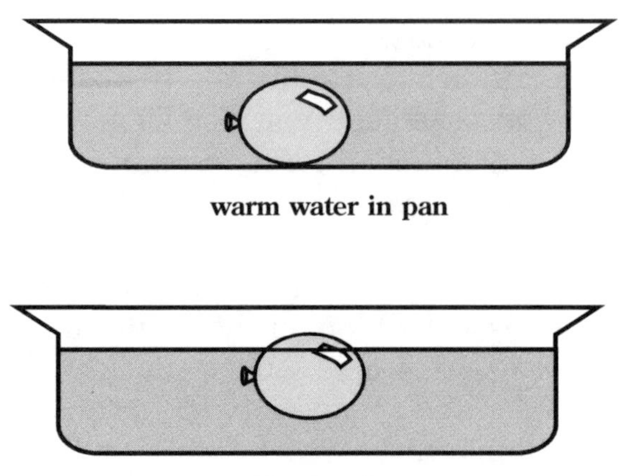

warm water in pan

cold water in pan

water dishpan for 10 minutes, take it out and place it in the warm-water dishpan. Does this balloon (which is now holding cold water) sink faster in warm water than the room-temperature balloon did?

Gases such as air are much less dense than liquids. It takes very little gas to dramatically affect the density of a water balloon. How much air can be added to the cold-water balloon before it is no longer able to sink to the bottom of the warm-water dishpan?

Floating and Sinking Eggs

You will need: 2 raw eggs, 2 clear drinking glasses, water, salt (preferably kosher salt), food coloring (blue and green), spoon

What to do: Fill two clear drinking glasses about two-thirds full with water. Stir salt into one of the glasses until no more salt dissolves. Add an egg to each glass. The egg placed in the glass of plain water will sink, but the egg placed in the salt water will float!

How it works: An egg is more dense than plain water, but less dense than salt water. So an egg sinks in plain water and floats in salt water. A substance's density, compared to water's can also be expressed as specific gravity. Specific gravity is the ratio of the density (weight per unit volume) of one substance compared to the density of water. Water's specific gravity is set at 1.00.

More science fun: Add blue food coloring to a glass of salt water and green food coloring to a glass of (unsalted) water. Place an egg in each glass and show the glasses to your family and friends. Fool them by telling them that eggs float in blue water but not in green water! Since you need a clear (not cloudy) glass of salt water to really fool your family and friends, it is best to use a salt that has no additives. Kosher salt is pure salt, and it will make this science trick work best. Kosher salt, unlike other kinds of salt, does not have any additives; it gives a salt water solution that looks as clear as unsalted water.

plain water

salt water

Liquids: Mixtures and Solutions

Dissolving Solids

You will need: sugar (powdered, granulated, and cubed), salt (regular iodized or noniodized, and kosher), 3 drinking glasses, water (cold, warm, and hot), spoon

What to do: Add a spoonful of powdered sugar to a drinking glass nearly full of water. Repeat the experiment with a second glass, using granulated sugar. Repeat the experiment in a third glass, using sugar cubes. Which form of sugar dissolves the most slowly? Which form dissolves the most rapidly? Take two fresh glasses of water. Dissolve regular (either iodized or noniodized) salt and kosher salt in the glasses of water. Which type of salt dissolves the best? Is one of the two salt solutions cloudy?

How it works: The rate that different forms of sugar dissolve depends upon the size of the sugar particles. The smaller the particles, the more quickly they dissolve. Regular salt gives a cloudy solution, and kosher salt does not. This is because regular salt has additives that keep it free-flowing when poured. These additives are not very soluble in water. Kosher salt, which is purer than regular salt, does not have additives. Kosher salt takes a little longer to dissolve than regular salt because its crystals are larger. But after kosher salt does dissolve, it does so completely, with no cloudiness.

More chemistry fun: Dissolve granulated sugar in cold, warm and hot water in separate glasses. How many more spoonfuls of sugar dissolve in hot water than in cold water?

Gelatin Formation

You will need: Flavored gelatin or unflavored gelatin, water, large drinking glass or plastic food container, toothpick, spoon or other kitchen utensil (table knife or fork)

What to do: In a large drinking glass or plastic food container, mix 1 packet of unflavored gelatin with 2 cups of hot water until it is completely dissolved. Put the glass into the refrigerator and keep it there overnight. The liquid will form a squishy semisolid called a *gel.* This gel

can be carved into shapes with a spoon or other kitchen utensil such as a kitchen knife. Stick a toothpick or fork into the gel shape to pick it up.
How it works: When gelatin is dissolved in just the right amount of hot water, it turns into a semisolid after it is cooled. The reason dissolved semisolid gelatin does not return to its solid powder form is that water holds the tiny particles of gel tightly, preventing it from changing back to a genuine solid. Since gelatin when cooled doesn't remain a liquid, it becomes a semisolid. A semisolid is a substance that is neither entirely like a solid nor like a liquid.
More chemistry fun: Use different amounts of hot water to dissolve gelatin, then cool the solution. How much or how little water can you use, and still get a gel? What happens if the water is just lukewarm (not very hot)? Does cold water dissolve gelatin at all?

Greasy Spots and Soapy Solutions

You will need: 3 paper cups, water, laundry detergent, dish detergent, bar of soap, spoon, solid vegetable shortening (or lard, butter, cooking oil, or meat grease), paper towel or rag
What to do: Pour an equal amount of water into three paper cups. Add a spoonful of laundry detergent to cup #1. Add a spoonful of dish detergent to cup #2. Add a piece of bar soap to cup #3. Cut three small pieces from a paper towel (or rag). Place a small amount of solid vegetable shortening (or lard, butter, cooking oil, or meat grease) on your finger.

Rub your greasy finger onto one end of each of the three pieces. Dip the end of the first paper towel piece into cup #1, the second piece into cup #2, and the third piece into cup #3. Pull out the pieces after a minute and allow them to dry. Which liquid best removes the greasy spots from the paper towels?
How it works: Detergents and soaps help to dissolve fats and greases such as vegetable shortening. One part of the detergent or soap molecule is attracted to water, and the other part is attracted to oils and fats. The attraction that detergents and soaps have for both shortening and for water allows the greasy spot to dissolve in water and thus to be removed from the paper towel.
More chemistry fun: Try different brands of detergents to see which ones remove the shortening from the paper towel the most quickly. Try different types of bar soap to see which ones work the best. Do soaps with lots of skin softeners in them (such as Dove®) work as well as pure soaps (such as Ivory®)?

Powder Push-Dunk

You will need: cornstarch baby powder, bowl or drinking glass, water, dish detergent
What to do: Sprinkle baby powder onto the surface of a bowl or drinking glass of water. Add several drops of dish detergent. The baby powder will scatter away from the detergent!

How it works: Detergent (or any soap) breaks the surface tension or skin of water. When detergent drops dissolve in water, the skin of the water is torn. Water molecules would rather be in contact with themselves than with detergent. So they move away from the detergent drops. As water molecules move from the center of the glass to the sides, the baby powder moves too.
More chemistry fun: Which types of dish detergent cause baby powder to move toward the sides of the bowl most quickly? Try the experiment with some powdered spices, such as finely ground pepper or cinnamon. Do these powders move as quickly as baby powder when the dish detergent is added?

Classy Crystals

You will need: salt, sugar, Epsom salt (magnesium sulfate), water, heat-resistant clear plastic drinking glass, heat-resistant plastic measuring cup, spoon, string, pencil
What to do: Place several spoonfuls of table salt into a heat-resistant plastic measuring cup. Fill the cup three-quarters full with water. In a microwave oven, heat the cup of salt and water until the liquid starts to bubble. Remove the cup and stir the salt water with a spoon. If all the salt dissolves, add another spoonful of salt. Repeat the microwave heating process, removing the cup and stirring the salt until the solution is saturated. *Saturated* means that no further salt will dissolve; a small amount of salt will remain on the bottom of the glass after stirring. Quickly and carefully pour the hot salt water from the measuring cup into a heat-resistant clear plastic drinking glass. Make sure to leave the undissolved salt behind in the measuring cup. Tie a string to the middle of a pencil. Lay the pencil across

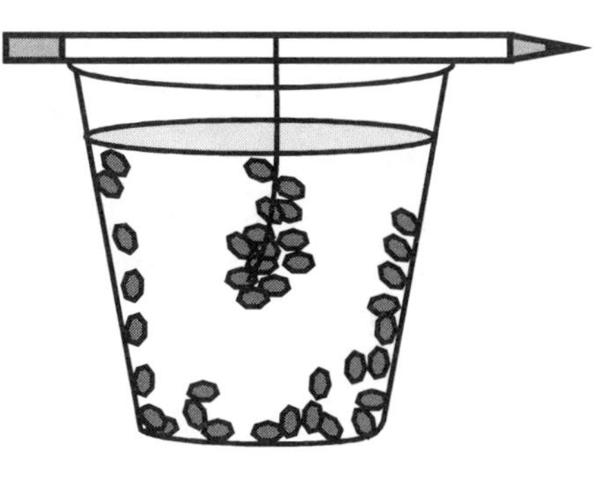

the center of the drinking glass containing the hot salt water solution. Use the spoon to move the free end of the string toward the bottom of the glass. Allow the glass to cool overnight. The next day, crystals will be attached to the string and also to the walls of the glass.

How it works: Solids such as salt dissolve better in hot water than in cold water. When you put as much salt into hot water as it can dissolve and then cool the mixture, some of the salt is forced out as crystals. These crystals form on any solid surface (string or walls of the glass) that is in contact with the liquid.

More chemistry fun: Repeat the experiment using sugar or Epsom salt (magnesium sulfate; available from a pharmacy) instead of salt. How do these solids compare with salt? How easily do they form crystals? Wait up to a week to make sure all of the crystals come out. Compare the shapes of crystals from salt (rectangular), sugar (irregular) or Epsom salt (beautiful slender rods). Try thread or yarn instead of string. Do crystals form as easily on thread or yarn? Are crystals on thread or yarn larger or smaller than crystals formed on string?

Stalactites and Stalagmites

You will need: string, yarn, or thread, 2 cups or glasses, plastic juice container, measuring cup, baking tray, water, Epsom salt, large spoon

What to do: Add one cup of Epsom salt to two cups of water in a plastic juice container and stir with a large spoon until the liquid is quite thick. Pour the solution into two cups or glasses that are close to each other (2 to 6 inches apart) in a baking tray (to catch any spills). Moisten an 18-inch piece of string with water. Place one end of the string into one of the cups of Epsom salt solution (below the surface) and the other end into the second cup. Leave enough slack in the string to allow it to droop below the height of the ends of the strings in the cups.

Let the experiment stand for several days. The solution of Epsom salt in water will rise out of the two cups into the string and travel through the string until the solution reaches the low point of the string between the two cups. Drops will form at this point and will drip slowly into the tray. As the dripping occurs, crystals will form in two places: from the lowest part of the string and on the tray, directly below the lowest part of the string. Crystals hanging down from the string are called stalactites; crystals rising up from the tray are called stalagmites. For best results, wait three days to a week. The first crystals to form will be stalactites. Stalagmites will form beneath the stalactites after several additional days.

How it works: The saturated solution of Epsom salt travels through the string by capillary action down to the lowest point in the string by siphon action. Capillary action is caused by the attraction that liquids have for interior solid surfaces in tiny tubes. Siphon action is

caused by the weight of the liquid in the string; gravity pulls the liquid along the string until it reaches the lowest possible point. After the Epsom salt solution is drawn up into the string leading out of the cup, it travels downward into that part of the string hanging between the cups. From each cup, capillary action first pulls the liquid into the string; then up to the bend in the string at the top of the cup. Siphoning (the pull of gravity on the liquid in the string) causes the liquid to continue moving until it comes to the lowest part of the string, where it drips onto the bottom of the baking tray.
More chemistry fun: What works better: cotton string, cotton thread, or cotton yarn? Does synthetic string, thread, or yarn work as well as cotton string, thread, or yarn? Create colored stalactites and stalagmites by putting food coloring into an Epsom salt solution. Put yellow food coloring in one cup and blue food coloring in the other. You will get green stalactite and stalagmite crystals!

Acids and Bases

Acids and Bases in the Kitchen

You will need: vinegar, lemon juice, orange juice, apple juice, sour salt (citric acid), carbonated beverage (soda pop), baking soda, laundry detergent, household ammonia, cup or drinking glasses

What to do: Place a spoonful of baking soda into a cup or drinking glass. Pour some vinegar into the glass. The baking soda will form a huge amount of foam! See if other household liquids will foam when mixed with bak-

ing soda. Try lemon juice, orange juice, apple juice, any carbonated beverage (soda pop), laundry detergent, and household ammonia.

How it works: The first four liquids (lemon juice, orange

juice, apple juice, and soda pop) are acids and will react with baking soda. The last two liquids (laundry detergent, and household ammonia) are bases and will not react.

Acids contain a positively charged hydrogen ion (H^+). Bases contain a negatively charged hydroxide ion (OH^-). When acids and bases are mixed together, they form a salt plus water (H_2O). When an acid reacts with the base baking soda, a salt plus carbonic acid (H_2CO_3) is produced. Carbonic acid is unstable and falls apart into water (H_2O) and carbon dioxide (CO_2). Carbon dioxide is the gas in the foam that is produced when acids are mixed with baking soda.

More chemistry fun: Mix a drop of the base laundry detergent (or household ammonia) with any liquid acid (vinegar, the fruit juices, and soda pop). Pour a small quantity of this detergent/acid mixture to a spoonful of baking soda. Does the baking soda foam? If it does, this shows that the solution remains acidic. Add more detergent to the detergent/acid mixture. Continue testing its acidic strength by mixing a small amount with baking soda, looking for foaming, and adding more detergent. When the detergent/acid mixture no longer causes baking soda to foam, the detergent has neutralized the acid.

Fruit Juices and Sodas

You will need: lemon juice, orange juice, apple juice, grapefruit juice, grape juice, pineapple juice, tomato juice, carbonated beverages (cola and non-cola), baking soda, paper or Styrofoam cups

What to do: Almost all fruit juices are acids. Carbonated beverages are acids, too, since they contain carbon dioxide (which combines with water to form carbonic

acid, a weak acid). Test some juices and sodas (cola and non-cola) for their acid strength by pouring each one onto baking soda in a cup.

How it works: The juice that makes the most and longest lasting bubbles is the strongest acid. Sodas are weaker acids than fruit juices. Therefore they will give fewer bubbles when added to baking soda. Acids have positive hydrogen ions (H^+), which react with baking soda's negative bicarbonate (HCO_3^-) ions. The stronger the acid, the greater the number of hydrogen ions. And the greater the number of hydrogen ions, the more carbonic acid (H_2CO_3) that is produced. Carbonic acid slowly breaks down into water (H_2O) plus carbon dioxide (CO_2). Carbon dioxide produces the foam seen in acid/baking soda neutralizations.

More chemistry fun: Try aging the fruit juices and sodas by letting them stand around for several hours to several days, and then combining them with baking soda. Which liquids lose their acidity the most quickly? Microwave soda in a paper or Styrofoam cup briefly (no more than 30 seconds). Then test it. Does a warmed-up cup of soda have as much acidity as a non-warmed (cooler) cup of soda?

Prepackaged Acid–Base Mixtures

You will need: baking powder, baking soda, antacid tablets, dinner plate, water, spoon, cups, cream of tartar

What to do: Add a spoonful of baking powder to a dinner plate. Place a spoonful of baking soda next to it. Pour a little bit of water onto both solids. The baking powder will fizz, but the baking soda will not. An example of a

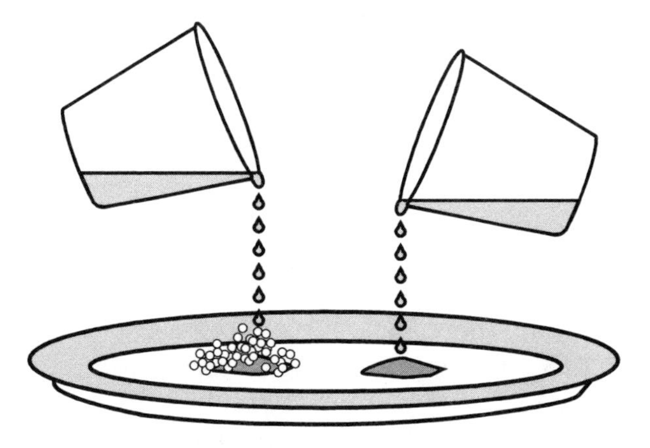

prepackaged ready-made acid-base mixture is an antacid tablet, such as Alka-Seltzer. Place one of these tablets on another dinner plate. Add water to the tablet. It will fizz, just as the baking powder did.

How it works: Baking powder is a mixture of baking soda (a base), an acid (such as calcium acid phosphate and/or sodium aluminum sulfate), and cornstarch (an extender/anti-caking ingredient). When water is added to baking powder, the base (baking soda) reacts with the acid (calcium acid phosphate) and bubbles of carbon dioxide form. Alka-Seltzer tablets (baking soda + citric acid) form bubbles in the same way: when the acid and base in the tablet dissolve in water, they react with each other, giving off carbon dioxide gas.

More chemistry fun: Try different brands of baking powder. Which work the best? You can create your own acid-base mixture by mixing sour salt (citric acid) with baking soda. When you add water to a mixture of sour salt and baking soda, it will fizz even more than baking powder! An opened can of baking powder that has been sitting around for months will not work as well as fresh baking powder. This is because water vapor in air dissolves small amounts of the baking powder's acid and base ingredients, which then neutralize each other. Another solid acid you can premix with baking soda is cream of tartar, derived from grapes.

How to Speed Up a Chemical Reaction

You will need: 3 Alka-Seltzer® tablets or other antacid tablets, baking powder, and a homemade acid–base fizzy mixture (baking soda + sour salt), water, 3 drinking glasses

What to do: Prepare three glasses of water at different temperatures. Add cold water and an ice cube to glass #1. Add lukewarm water to glass #2. Add hot water to glass #3. Quickly add an antacid tablet to each of the glasses. What do you see happening in each glass of water?

How it works: Each glass of water gives off bubbles after an antacid tablet is added. Hot water produces bubbles fastest, lukewarm water more slowly, and ice water most slowly of all. An antacid tablet reacts with itself when it dissolves in water. The active fizzing ingredients in the tablets are citric acid (an acid; also known as sour salt) and sodium bicarbonate (a base; also known as baking soda). The medicine in Alka-Seltzer tablets is aspirin, but it does not cause the tablet to fizz. Baking soda contains carbon dioxide, which is released as a gas

when baking soda is neutralized with acid. Hot water speeds up an antacid tablet's fizzing action in two ways. It (1) quickly dissolves the tablet, and (2) speeds up the reaction of the acid and base in the tablet. Hot water in this experiment gives a very dramatic effect, causing the antacid tablet to noisily, almost explosively, dissolve in just a few seconds!

More chemistry fun: Do the temperature experiment (using hot, warm, and cool water) with baking powder. Do it again with a homemade acid–base fizzy mixture (baking soda + sour salt). Which reacts faster: baking powder or your homemade acid–base fizzy mixture?

Balloon Self-Inflator

You will need: large round balloon, 2-liter plastic soda bottle, vinegar, funnel, waxed paper, baking soda, spoon

What to do: Cut a 10 inch x 6 inch piece of waxed paper. Spoon a half-inch-thick strip of baking soda powder down the center of the long (10-inch) dimension of the paper. Roll the waxed paper to give a 10 inch long, half-inch-thick cylinder. Fold over the two ends. Push the cylinder into a 2-liter plastic soda bottle. Pour 1 to 2 cups of vinegar into the bottle through a funnel. Quickly

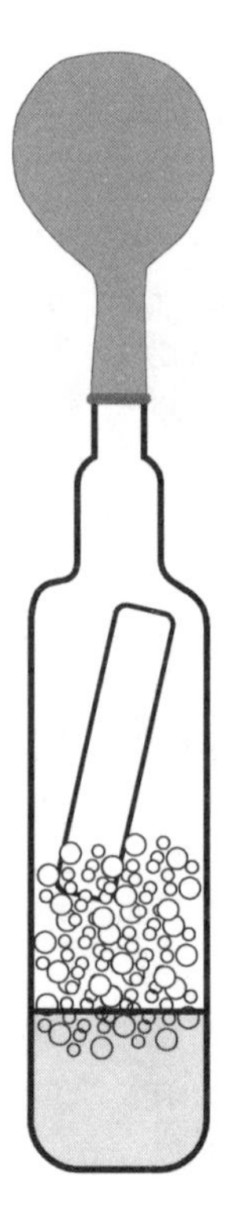

stretch the mouth of a large round balloon over the end of the bottle. Pick up the bottle and shake it vigorously for 10 to 20 seconds. The balloon will rapidly swell up!

How it works: When you shake the bottle, a small amount of vinegar moves to the inside of the cylinder of baking soda. When it does, the two (vinegar and baking soda) react and produce carbon dioxide gas. This gas release causes the cylinder to puff up, which allows even more vinegar to move into the cylinder to react with the baking soda. The gas pressure inside the bottle quickly increases, inflating the balloon.

More chemistry fun: Add the baking soda in a folded-up paper towel instead of waxed paper. How quickly, and how much, does the balloon blow up, compared with when baking soda was added in waxed paper? Try using vinegar at different temperatures (ice-cold, room temperature, and warm). Which vinegar temperature causes the balloon to inflate the most?

Purple Cabbage Juice Indicator

You will need: purple cabbage, warm water (or rubbing alcohol), resealable zipper closure plastic bag, acids (white vinegar, lemon juice, and other fruit juices), bases (baking soda, detergent, and milk of magnesia), cheese grater, cups, microwave oven

What to do: Shred pieces of purple cabbage with a cheese grater. Put the pieces in a resealable plastic bag. Add water to half-fill the bag. Heat the bag containing the cabbage and water in a microwave oven for 20 to 30 seconds until the water becomes warm to hot, and the water turns dark purple. Shake the bag. Pour the purple cabbage juice into a cup, leaving the cabbage pieces in the bag. The purple liquid is an acid–base indicator, since it will change to different colors when it is added to an acid or a base. Mix it with acids, such as white vinegar, lemon juice and other fruit juices. The purple color will change to pink! Mix it with bases, such as baking soda, detergent, and milk of magnesia. The purple color will change to green!

How it works: Purple cabbage juice, when in water (H_2O), does not have a charge on it. When exposed to an acid (H^+), the juice turns into a positively charged species, which is pink. A *species* in acid–base chemistry is usually a molecule that has a positive or negative charge. When exposed to a base (OH^-), the juice turns into a negatively charged species, which is green.

More chemistry fun: Instead of extracting the purple color with warm water, shake the cabbage pieces with rubbing alcohol at room temperature. Caution: *do not heat!* Rubbing alcohol is flammable! Which is a better indicator (gives more dramatic color changes): purple cabbage juice in warm water or in rubbing alcohol?

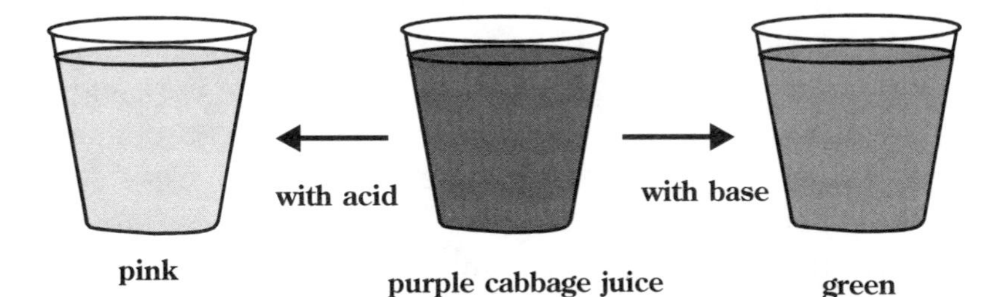

More Food Acid–Base Indicators

You will need some of the following: whole vegetables and fruits (beets, blueberries, cantaloupes, carrots, cherries, cranberries, grapes or grape juice, kiwifruit, mangos, red and yellow onions, papayas, spinach, strawberries, sweet potatoes, watermelons), skins of fruits and vegetables (apples, apricots, nectarines, peaches, pears, plums, radishes, rhubarb, and turnips), colored spices (such as cumin, curry, and turmeric), colored ice pops, colored juice drink, colored gelatin, warm water, resealable zipper closure plastic bag, acids such as lemon juice and other fruit juices, bases such as baking soda and detergent, cheese grater, and cups, microwave oven

What to do: Many foods other than purple cabbage will undergo color changes when mixed with acids or bases. Selecting from the foods listed above, chop up or shred the food with a cheese grater. Heat the food up briefly in a resealable plastic bag just as you did for the purple cabbage juice experiment (page 46). Pour off the colored liquid extract into a cup. Try different acids and bases found in the kitchen to see if the extracts make good indicators, and, if so, how many colors they can make when mixed with different acids and bases. Which of the indicators gives color changes similar to purple cabbage juice indicator? Which indicators give brand-new color changes?

How it works: Just as with purple cabbage juice, other food indicators when suspended or dissolved in water do not usually have a charge on them. When exposed to an acid (H^+) or base (OH^-), many food indicators gain a positive (+) or negative (-) charge. These charged food indicators often have colors very different from the uncharged form. Many indicators (including purple cabbage juice) can actually go through more than two color changes when exposed to acids and bases. The more complex and brightly colored the substance which is an indicator, the greater the number of color changes, and the easier it is to see those colors.

More chemistry fun: Instead of extracting the colored fruit or vegetable with warm water, use rubbing alcohol at room temperature. Caution: rubbing alcohol is flammable; do not heat! Which liquid is a better solvent for food indicators: warm water or rubbing alcohol?

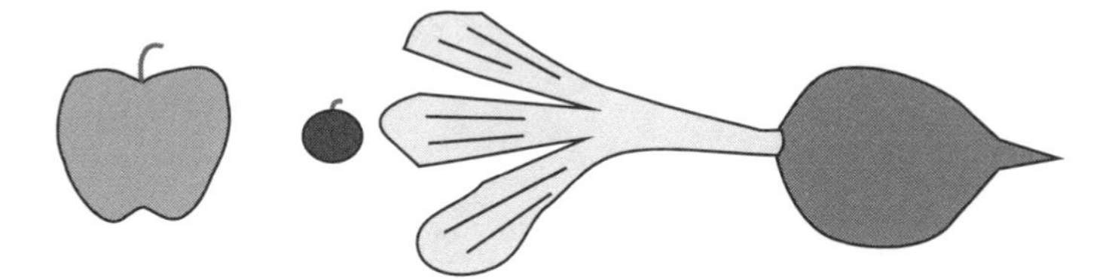

Vinegar Volcano

You will need: vinegar (distilled or clear), waxed paper, baking soda, spoon, water, red food coloring, liquid dish detergent (not laundry detergent), 2-liter plastic soda bottle, funnel, dirt pile, measuring cup

What to do: Do this outdoors, as it may make a mess. Place a 2-liter plastic soda bottle outdoors in a dirt pile. Push dirt around the bottle until only the neck is sticking out of the dirt pile. Wrap a couple of spoonfuls of baking soda in a thin tube of waxed paper (see page 45) and shove the tube into the mouth of the bottle. Stick a funnel into the mouth of the bottle. Add a spoonful of a liquid dish detergent. Add a few drops of red food coloring. Add 1 liter (about 4 cups) of water. Now quickly add 1 to 2 cups of vinegar. The bottle will erupt with red

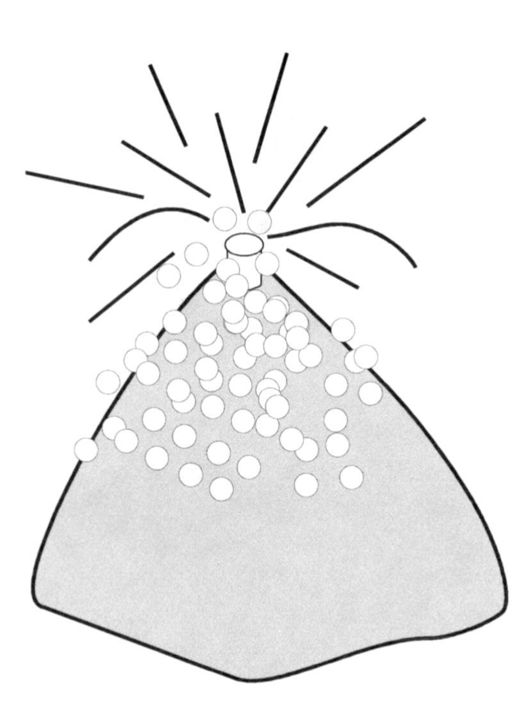

foam as soon as the vinegar soaks through the waxed paper and reacts with the baking soda!

How it works: Vinegar reacts with baking soda to form carbon dioxide. The dish detergent considerably increases the amount of foaming. The red food coloring imitates the color of a real volcano. The reason to use a liquid dish detergent and not a laundry detergent is because dish detergents are designed to foam a lot. Laundry detergents are not supposed to foam. Automatic dishwasher detergents aren't supposed to foam either.

More chemistry fun: Try using water and vinegar at different temperatures (cold, warm, and hot). Does a vinegar volcano work better (give more foam) with hot water than with warm or with cold water? Put some drops of yellow food coloring around the base of the volcano. Add blue food coloring to the vinegar before adding it to the bottle. When the volcano erupts, the yellow drops will turn green when the vinegar/baking soda/water foam touches them!

Vinegar–Baking Soda Foam Sprayer

You will need: vinegar, waxed paper, baking soda, spoon, funnel, water, liquid dish detergent, 2-liter plastic soda bottle, drinking straw, clay

What to do: Do this experiment over the sink, as it may get messy. Wrap a couple of spoonfuls of baking soda in waxed paper (see page 45) and shove the paper into a 2-liter plastic soda bottle. Add a spoonful of a liquid dish detergent. Fill the bottle half-full with water. Shape a piece of clay around a drinking straw so that most of the straw will stick into the bottle when it is placed on the bottle, but don't put the straw on yet. Through a funnel, add a cup of vinegar to the bottle. Quickly stick the clay-wrapped straw onto the mouth of the bottle. Shake the bottle vigorously to dislodge the baking soda from its waxed paper container. Liquid and/or foam will spray out of the straw!

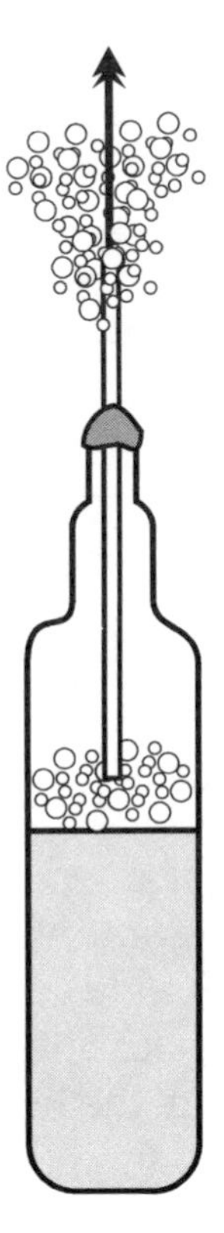

How it works: Vinegar reacts with baking soda to form carbon dioxide gas. Dish detergent greatly increases the amount of foam. If the bottom part of the straw is below the liquid level in the bottle, liquid will shoot out of the straw! If the bottom part of the straw is above the liquid level, foam will shoot out of the straw!

More chemistry fun: Try different ratios (proportions) of baking soda, vinegar, dish detergent, and water. Which ratio gives the best (farthest-squirting) foam sprayer?

Food Color Foamy Fountain

You will need: distilled white vinegar, baking soda, liquid dish detergent, funnel, dish pan or pie pan, 3-oz paper cup, food coloring (blue and yellow)

What to do: Place a mound of baking soda in a dish pan or pie pan. Pour a spoonful of a liquid dish detergent onto the mound of baking soda. Place a funnel upside-down over the top of the mound. Using a 3-oz paper cup (pinched on one side to give a narrow pouring spout),

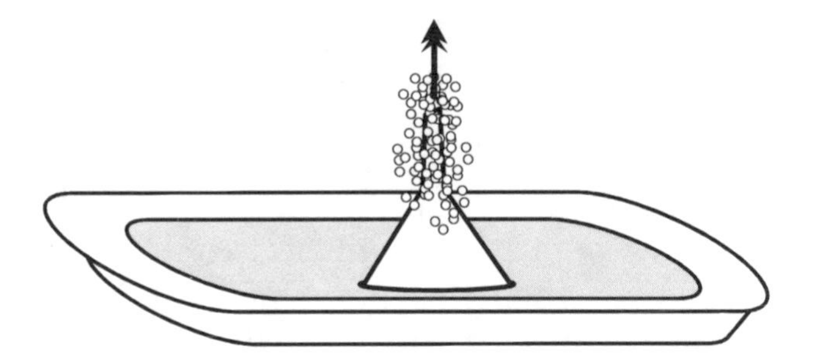

quickly pour vinegar through the neck of the funnel. Foam will spurt out of the funnel's neck, and some also around the funnel's rim (against the dishpan).

How it works: Vinegar reacts with baking soda to form carbon dioxide gas. Dish detergent greatly increases the amount of foam. The gas pressure of the carbon dioxide causes foam to rush out up through the funnel's neck. The more tightly you hold the funnel against the pan, the more foam will come out of the funnel neck opening (and the less foam will leak out around the rim).

More chemistry fun: Have fun with color changes! Put a few drops of yellow food coloring onto the baking soda and add some blue food coloring to the vinegar. When you activate the volcano by mixing the vinegar with the baking soda, you will get green foam!

Food Chemistry

Air Browns an Apple

You will need: apples, lemon juice, knife, paper towels
What to do: Slice each apple in half with a kitchen knife. Rub one of the halves with a lemon juice-soaked paper towel. Leave the other one alone. Let the two halves stand in air for 5 to 10 minutes. Which half browns the

no lemon juice **lemon juice**

faster? Try the experiment in a refrigerator. Does the brown color appear as quickly when the apple halves are kept cold?
How it works: Air contains oxygen. Oxygen reacts with the exposed flesh of a cut-open apple. Air-reacted apple flesh has a brown color. This browning can be prevented or slowed down by protecting the exposed fruit with an antioxidant, something that slows down the joining of the oxygen with the apple. Lemon juice is a naturally occurring food antioxidant. Lemon juice, when applied to freshly cut fruit, greatly reduces the rate that browning (oxidation) occurs. Temperature also has an effect on oxidation. When apple halves are placed in the

refrigerator, they do not turn brown as quickly, because chemical reactions (such as the browning of an apple) slow down at low temperatures.

More chemistry fun: Protect apple slices with other citrus juices (orange, lime, and grapefruit), and also with vinegar. Do these liquids prevent apple slices from turning brown as well as lemon juice does? Compare fresh lemon juice with bottled lemon juice. Which one works better to prevent fruit browning? Try air-oxidizing other fruits instead of apples. Do other fruits when exposed to air turn brown as quickly as apples? Are they protected as effectively by lemon juice as apples are?

Milk Curdling

You will need: milk (skim, 1%, 2%, and regular 4%), 3 clear drinking glasses, acidic liquids (vinegar, lemon juice, and pineapple juice), spoon

What to do: Pour a small volume of regular (4%) milk into each of three clear drinking glasses, so each is less

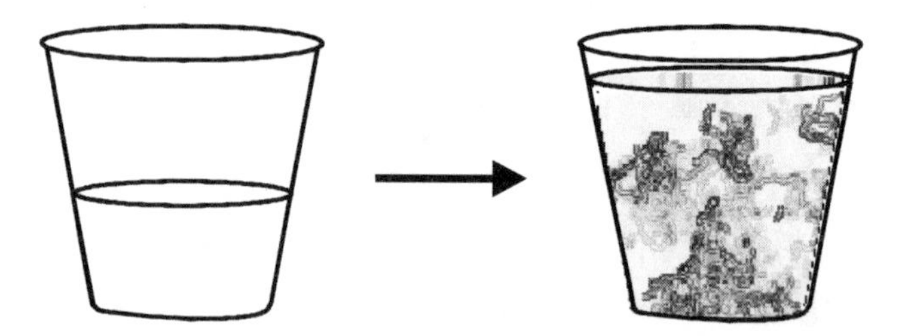

than one-third full. Add an equal volume of the following: vinegar to cup #1, lemon juice to cup #2, and pineapple juice to cup #3. Stir each cup vigorously with a spoon. Let the milk + acidic liquid mixtures stand for a few minutes. The mixtures will separate into a fine white solid and a watery pale milk-colored liquid.

How it works: Milk is a complex mixture of proteins, fats, and minerals in water. The protein part of milk consists of a suspension of invisible finely divided particles spread evenly throughout the liquid. Such a suspension is called a *colloid.* When an acidic liquid such as vinegar is added to milk, particles of protein and fat pull toward each other and are changed into a solid called curds, which separate from the liquid. The watery pale milk-colored liquid from which the curds separate is called whey. To obtain a sample of the curds, pour the cup of curds and whey onto a paper towel. The towel will absorb the liquid whey, leaving behind the solid curds. Forming curds and whey from milk is called curdling.

More chemistry fun: Compare milk having different amounts of fat in it (skim, 1%, 2%, and regular 4%). Which type of milk gives the most curds when mixed with vinegar? Warm a cup of milk by microwaving it for 30 seconds. When vinegar is added to the warm milk, does curdling occur faster?

Sugar: How Sweet It Is!

You will need: cane sugar (granulated, powdered, brown, and cube), honey, soda pop (cola and non-cola), diet soda, fruit juice (apple and orange), cookies, small paper cups, water, spoons

What to do: Add a spoonful of powdered sugar to a small paper cup filled with water. Add granulated sugar to a second paper cup of water. Place a sugar cube into a third cup of water. Stir the cups with spoons until the sugar dissolves. Which form of sugar dissolves the most quickly? Which form dissolves the most slowly?

Offer different forms and kinds of sugar to insects (such as ants or flies). Do insects prefer granulated, powdered, brown, or cube sugar? Solid sugar or sugar dissolved in water? Honey or sugar water? Sugared cola, sugared non-cola, or diet cola/non-cola? Fruit juice or sugared soda pop?

How it works: Finely divided sugar has more surface area than coarse sugar does. The finer (tinier) the sugar's particle size, the more quickly it dissolves in water. A sugar cube has very little surface area compared with granulated or powdered sugar. Therefore a sugar cube takes a long time to dissolve. Sugar-loving insects (such as ants and flies) are attracted more to solutions of sugar than they are to solid sugar. This is because they can smell sugar better in solution than as a solid, and also because they can more easily drink a liquid than eat a solid. Insects prefer sugared soft drinks over sugar-free, diet sodas. This is because they use sugar as a food. Artificial sweeteners, no matter how sweet they taste to us humans, do not have any value as a food. Bugs aren't so dumb after all!

More chemistry fun: Offer other sugar-containing foods such as fruit juices (both sweet and sour) and cookies (both sugar-coated and non-sugar-coated) to insects. Do insects prefer sweet fruit juices or sour fruit juices? Do they prefer sugar-coated cookies or non-sugar-coated cookies?

Salt: Kosher and Other Types

You will need: salt (regular, iodized, and kosher), sugar, 3 clear drinking glasses, spoons, water

What to do: Stir a spoonful of one of the following into each glass: regular, iodized, and kosher salt. Add water to each glass until it is half-full. Stir each glass with a spoon. Notice how quickly each type of salt dissolves. Do all three types of salt give clear solutions? Which ones do not?

How it works: Kosher salt is the only one of the three types of salt in this experiment that gives a clear solution. The other salts (regular and iodized) give cloudy solutions. Because of its larger crystal size, kosher salt takes longer to dissolve than regular or iodized salt, but when it finally does dissolve, it does so completely. Salt (sodium chloride) is chemically made up of two elements: sodium and chlorine. Regular and iodized salt, unlike kosher salt, have extra ingredients in them. These ingredients give salt desirable properties. Salt when exposed to air picks up moisture, which causes its crystals to stick together. To stop the crystals from sticking, an anticaking agent (sodium silicoaluminate) is added. Dextrose, a sugar, is often added to absorb moisture and to make the salt taste slightly sweet. Iodized salt has potassium iodide in it, which is an essential chemical in the human diet.

More chemistry fun: Look closely at salt crystals and compare them with sugar crystals. You will see that the crystals have different shapes. Salt is square-looking; sugar crystals are irregular. Do you notice any difference (besides size) between regular salt (with additives) and kosher salt? Mix a spoonful of regular salt with a spoonful of granulated sugar on a dinner plate. How easily can you separate the crystals? Try using a toothpick to move the crystals apart.

Shining Pennies

You will need: dull pennies (or any copper-containing coins), table salt, vinegar, paper towels, spoon

What to do: Place a dull penny (or any copper-containing coin) on a paper towel. Sprinkle some table salt on the penny. With a spoon, pour a little vinegar on top of the salt. Rub the penny. It will immediately turn shiny!

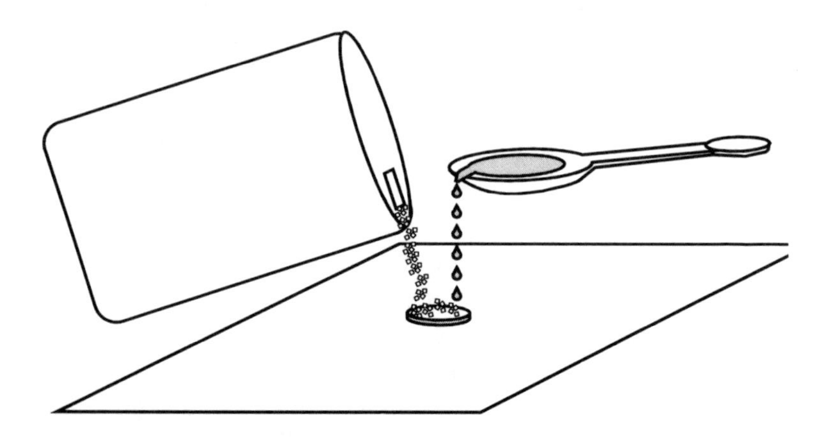

How it works: Table salt is sodium chloride. Vinegar is acetic acid in water. The chloride ions in table salt combine with the hydrogen ions in vinegar to produce a small amount of a strong acid (hydrochloric acid), which eats off a little bit of the penny's surface. In addition, salt is somewhat abrasive (rough like sandpaper). When you rub salt on the penny, the abrasive quality of salt helps the acid to remove the dull surface from the penny, exposing the hidden shiny surface.
More chemistry fun: Do this experiment with: (a) salt only; (b) vinegar only; (c) lemon juice only; (d) salt and lemon juice. Does any of these treatments work as well as salt and vinegar does?

Food Color Explosion

You will need: baking dish (or bowl) of milk that has 1% to 4% fat content, food coloring (red, yellow, green, and blue), dishwashing liquid, cornstarch-based baby powder, powdered pepper
What to do: Half-fill a baking dish (or bowl) of milk that has 1% to 4% fat content. Add eight drops of food color-

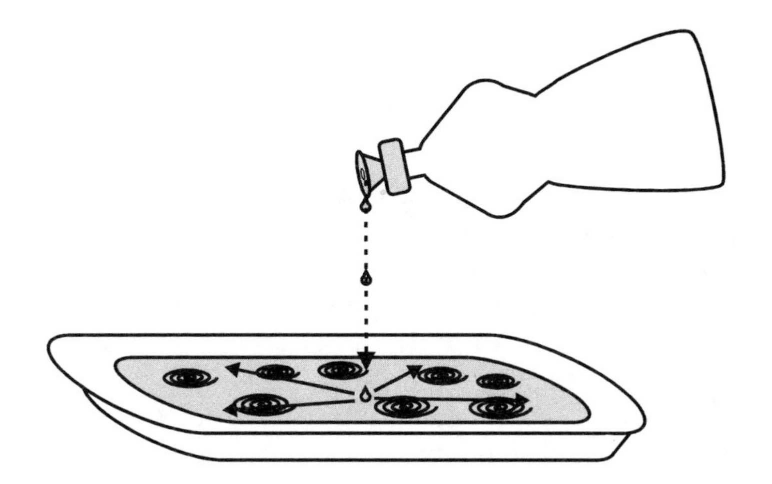

ing to the dish of milk near the center of the dish (two drops each of red, yellow, green, and blue). Add two or three drops of dishwashing liquid to the surface of the milk. Watch the food colors explode and swirl around in beautiful patterns!

How it works: Milk fat holds the droplets of food coloring in place because milk fat and the water in food coloring do not like to mix or to dissolve each other. As soon as dishwashing liquid (or any soap or detergent) is added, the food coloring drops and the milk rapidly mix together. Dishwashing liquid (or any soap or detergent) is made of molecules that have two ends: a fat-friendly end, and a water-friendly end. One end of the dishwashing liquid attracts the fat in milk, and the other end attracts the water in the food coloring. The food coloring mixes with the milk because the dishwashing liquid acts as a bridge between the two.

More chemistry fun: Which works better: milk that has a low (1%) fat content or a high (4%) fat content? Use baby powder or powdered pepper instead of food coloring. Do these powders swirl around the dish like food colors do? How many dishwashing liquid drops are needed to give the best swirling action of food colors, of baby powder, or of powdered pepper?

Puffed-Up Rice

You will need: dry (uncooked) rice, 2 clear drinking glasses, water, tape, pencil or marker, ruler, beans

What to do: Fill a clear drinking glass half-full with water. Apply tape vertically (up and down) on the outside of the glass. Mark the water level with a pencil or marker and measure it with a ruler. Write down the height of the water. Add rice to another glass of the same size until it is about one-fourth full. Measure it with a ruler. Write down how high the rice is in the glass. Add the rice in the second glass to the water in the first glass. Does the water level rise? Does it rise as much as you thought it would?

How it works: Although the water level does rise when a glass of water and rice are mixed, the level does not rise as high as the total of the heights of the water and rice in their separate glasses. A glass of dry rice has air hidden between the rice grains. When rice is poured

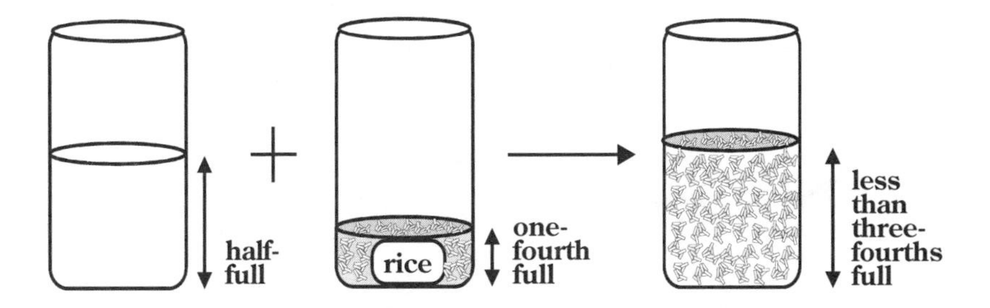

into the glass of water, the hidden air is displaced (pushed aside). Since you can't see this air, it looks like the volume of rice and water after mixing is less than the volume of the rice and water before mixing.

More chemistry fun: Try different kinds of rice (short and long grain white, instant rice, and brown rice) to compare how much they raise the height of water. Also try different kinds of beans (pinto, kidney, navy, and lentils).

Colors Creep Up Celery Stalk

You will need: fresh stalk of celery, red food coloring, large drinking glass, water, plastic table knife, beets, red cabbage

What to do: Cut the bottom from a fresh stalk of celery with a plastic table knife. Add red food coloring to a large glass about one-fifth full of water. Place the bottom end of the celery in the glass. Look at the celery stalk after it has stood for several hours. The lower part of the stalk will have a pale pinkish color. Continue to check on the stalk daily for three days. After this period of time, the stem and the leaves on the top will be colored a pinkish to pale red color! Look down on the top of the stalk to see the red color in the capillaries (the tiny tubes in the stem).

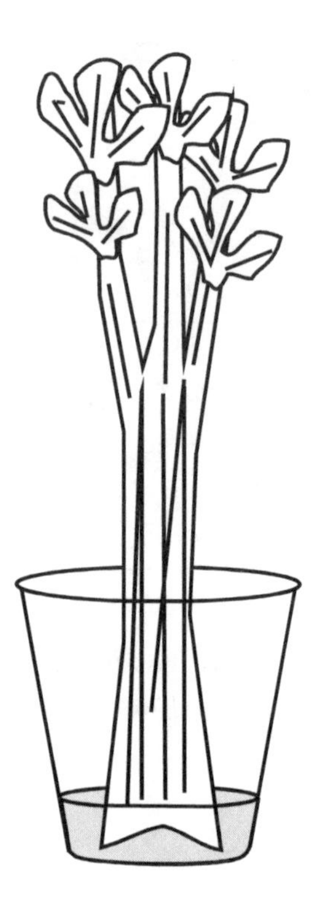

How it works: Colored water travels up into the celery stalk by a scientific principle called capillary action. Capillary action is the tendency of liquids in tiny tubes to rise in the tubes. This happens because air pressure pressing on the liquid in the glass (and thus on the bottom of the celery stalk) is greater than the air pressure inside the stalk. The stalks and stems of most plants have capillaries in them to carry water from their roots up to their leaves.

More chemistry fun: Add the juice from a colored food such as beets or purple cabbage to a glass with a little bit of water in it. Does beet or red cabbage color travel up a celery stem as fast as food coloring does? Are the leaves as noticeably colored?

Coffee Filter Food and Candy Chromatography

You will need: cone style coffee filter, fruits and vegetables with intense colors (such as beets, purple cabbage, and dark cherries), candies with colored sugar shells, dinner plate, water, food grater, watertight zipper closure sandwich bag

What to do: Grate each colored fruit and vegetable into tiny pieces with a food grater. Place the chopped bits of each food into its own watertight sandwich bag. Add a

small amount of rubbing alcohol. Allow the bag to stand for a few moments until the rubbing alcohol is deeply colored. Turn a cone style filter upside down and place it on a plate. Spoon the food-colored liquids in thin bands of color one-half inch away from and parallel to the mouth of the filter. Create another thin strip of color with a rubbing alcohol extract of a candy that has colored coatings. Add a small amount of water to the plate, enough to touch all parts of the mouth of the filter, but not touching the colored marks. As water rises in the filter, the colored marks will move up too, separating into new colors! For example, red beet extract will separate into a fast-moving pale orange band with a larger red band just behind it.

How it works: Water rises in the filter by capillary action. As it does so, the food, vegetable, and candy colors on the coffee filter dissolve into the water and move up the filter. Any colors that are mixtures of more than one color can be separated on the filter. Those colors which are hydrophilic (water-liking) will move quickly, moving nearly as fast as the water does. Those colors which are hydrophobic (water-disliking) will move more slowly. Compare the intensity on the filter of food colors with candy colors. Which are more brightly colored?

More science fun: Create different patterns and thicknesses of marks. Try to move the spots with solvents other than water, such as salt water or rubbing alcohol. Do colors move faster in rubbing alcohol or in water? Instead of fruits and vegetables, try extracts from brightly colored (red and yellow) tree leaves and flowers.

Yeast–Sugar Balloon Inflation

You will need: packet of active dry yeast, sugar, warm water, large round balloon, 1-liter plastic soda bottle, funnel, tablespoon

What to do: Place a funnel in the neck of a 1-liter plastic soda bottle. Quickly add through the funnel: (1) the contents of a packet of dry yeast; (2) a tablespoon of sugar; and (3) a cup of lukewarm water (temperature = 105 to 115 degrees Fahrenheit, just above your body temperature of 98.6 degrees). When all three ingredients are added, pull the end of a large round balloon over the mouth of the soda bottle. Make sure to stretch the neck of the balloon completely over the threads of the bottle mouth. Otherwise, the gas produced in the bottle will leak out and the balloon will not inflate. After you are sure that the balloon is tightly connected to the mouth of the bottle, shake the bottle vigorously for 30 seconds. Let the bottle stand for 15 to 30 minutes. After this period of time, the balloon will puff up. It will continue to slowly puff up for the next two hours! If you hold your ear up against the bottle about 20 minutes after the experiment is started, you'll actually *hear* the yeast eating up the sugar (a fizzing sound from carbon dioxide gas being produced).

How it works: Yeast is a living plant that feeds on sugar. When mixed with sugar and water, yeast makes carbon dioxide gas as a by-product. The pressure from this gas puffs up the balloon. The foam you see on the surface of the liquid in the bottle is composed of tiny carbon dioxide bubbles being formed by yeast as it eats the sugar. Sniff the outside of the balloon an hour after the experiment is started. You will smell yeast! This is because some of the carbon dioxide diffuses (passes through) the balloon skin. As it does so, it carries the distinctive smell of yeast with it.

More chemistry fun: With the same experiment, try water at different temperatures (cold, warm, and hot). Which gives the best (quickest and fullest) balloon inflation? Also try different amounts and kinds of sugar (granulated, powdered, brown, and cube). Which type of sugar does yeast like the best (which causes the balloon to puff up the most)? Compare quick-rising yeast with regular yeast. Which yeast works more quickly? Which yeast causes the balloon to puff up the most?

Vinegar Eats an Eggshell

You will need: egg, distilled (clear) vinegar, lemon juice, drinking glass, plastic wrap

What to do: Carefully place an uncooked egg in the bottom of a drinking glass. Pour clear vinegar over it until it the egg is covered. Cover the top of the glass with plastic wrap (this helps keeps any bad smell from escaping). Tiny bubbles will begin to form on the eggshell right away. Look at the egg every couple of hours or so. After 18 to 30 hours (about a day's time), the shell will have been dissolved, leaving the raw egg in its membrane. You will be able to see into the egg. Little bits of egg may also come out of the membrane. Interestingly, after standing overnight, the egg (which had been resting in the bottom of the glass the previous day) will now be *floating!*

How it works: Vinegar is a solution of acetic acid in water. Acetic acid reacts with a basic chemical (called calcium carbonate) that is in the shell of the egg. This reaction produces carbon dioxide gas and calcium acetate. The bubbles that appear on the outside of the egg after vinegar is added are evidence that the eggshell-eating reaction has started. The reaction will continue until all of the eggshell has been removed. The calcium carbonate in the eggshell makes an egg more dense than vinegar. So the egg sinks to the bottom of the glass of vinegar at first. When the dense shell is gone, the rest of the egg (now less dense than vinegar) floats. Bubbles of carbon dioxide sticking to the outside of the egg also help the egg to float.

More chemistry fun: Experiment with vinegar at different temperatures: cold (in the refrigerator), moderate (room temperature), and warm (outside on a hot sunny day, with a dark-colored bag over the glass containing the egg). Use another mild kitchen acid (such as lemon juice) in place of the vinegar. Does lemon juice dissolve the eggshell as quickly as vinegar does? After the shell has been chemically removed with vinegar, place the egg on a dinner plate. Poke into the egg with a fork. The egg will burst and run onto the plate, exposing the whitish membrane that had been holding the egg together.

Chemical Detective

Invisible Inks and Incredible Messages

You will need: lemon juice, small paintbrush (or toothpick broken in half), white writing paper, paper cup (3-oz size)

The situation: You want to send a secret message to your best friend, but you want to keep your pesky brothers and silly sisters from reading it (or even from guessing that a message is written on it)!

What to do: First, write a normal message with a black-ink pen or marker on the upper half of a white sheet of paper (this will distract unauthorized readers from the real message). To prepare your ink for the second (secret) message, pour a small amount of lemon juice into a 3-ounce paper cup. Dip a small paintbrush or broken toothpick into the lemon juice. Write on the lower half of the paper. Allow the juice to dry. This will take about 10 minutes to an hour. Helpful hint: the less secret ink (lemon juice) you use, the faster it will dry. To help it dry faster, take the paper outside on a windy day and wave the paper in the wind. To dry the paper indoors, hold it in front of an electric fan. Send or give your paper to a friend. Tell your friend to hold it over a light bulb or other heat source, such as a toaster. Your secret message will quickly appear!

How it works: The sugars, acids, and other chemicals in lemon juice turn brown when heated, more so than paper does. Because of this, your secret dried lemon juice message is easily seen on white paper.

More chemistry fun: Try using other liquids for invisible inks, such as orange juice, grapefruit juice, apple juice, tomato juice, milk, sugar water, salt water, and detergent. Which liquids work the best? Do any of them work as well as lemon juice does?

Secret Ink Messages

You will need: pens and markers of various types and colors (mainly black), paper towel, 3 small dinner plates, water, salt, rubbing alcohol

The situation: In your job as a detective, you regularly need to send messages that contain hidden information. These messages will tell your friends where to meet, what time to meet, and so forth. The way to keep this information secret from strangers and non-friends is to choose a series of inks that look the same but behave differently when chromatographed. The process of spotting of a chemical (such as ink) on an absorbent material (such as a paper towel), dipping an edge of that material in a liquid (such as water), and seeing the spot separate into colored spots is called chromatography. The word *chromatography* comes from the Greek word for color.

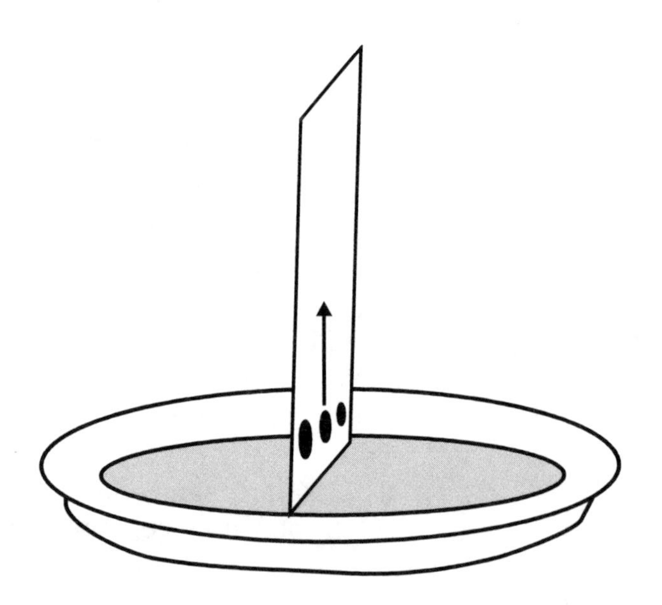

What to do: Select pens and markers of various types and colors. Place a spot of ink from each pen or marker about a half-inch from one edge of a paper towel. Put a small tear in the towel at the edge nearest where the spots are. Dip the torn edge of the paper towel into one of three liquids: water, salt water, or rubbing alcohol in a dinner plate. Hold the towel in place for a few minutes while the liquid moves up the towel. Notice what happens to the spots. Pen inks starting out the same color can behave quite differently from one another when they are applied to a paper towel and the towel dipped into a plate of liquid. For example, the black ink used to send your message could do one of three things: (1) stay black and not move; (2) stay black and move; (3) move, but turn into different colors as it goes up the paper towel. You could make a code in which each behavior could convey a different message. For example, (1) could mean "meet me at the flagpole"; (2) could mean "meet me at the bus stop"; and (3) could mean "meet me in the gym."

Experiment with waterproof and water-soluble markers and with water, salt water, and alcohol to see what results you get.

How it works: Water rises in the paper towel by capillary action. Under the right conditions, inks from the pens and markers separate into different colors on the paper towel. Many inks that look the same before this treatment look very different afterwards (either by moving at different rates, by separating into colors, or both). The chemical and physical properties of the inks cause them to hold onto the paper with varying degrees of strength when pushed with a solvent such as water.

More chemistry fun: Develop paper towels' pen and marker spots in other liquids, such as: (a) dilute household ammonia (hold your nose to avoid the bad smell!); (b) clear (distilled; white) vinegar; and (c) water with baking soda dissolved in it. This last solution (c) can be very impressive if you write secret messages with sour salt (citric acid) added to the writing ink, since sour salt foams when it touches baking soda in water! Another idea is to send a message written with fountain pen ink mixed with baking soda. Use a toothpick to write with. When your friend dips this paper towel into a plate of vinegar, the pen ink will bubble when the vinegar reaches it! The bubbling is caused by baking soda (in the ink line) reacting with vinegar to form the gas carbon dioxide.

Buzz-Fizz, the Evil Bug Emperor

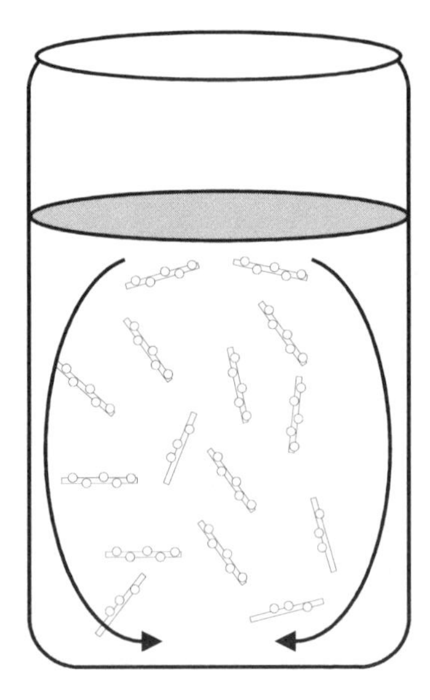

You will need: water, baking soda, white (distilled) vinegar, spoons (2), cups or drinking glasses (at least two), food such as uncooked beans, cherries, grapes, lentils, nuts, peas, popcorn (unpopped), raisins, seeds (apple, bird, and sunflower seeds), half-inch to 2-inch pieces of dry spaghetti; small wadded-up pieces of aluminum foil

The situation: Buzz-Fizz, the Evil Bug Emperor, is secretly plotting to steal treasure at the bottom of a lake filled with magic liquid. He plans to send pieces of food to do his dirty work for him. These pieces of food have to have some very special skills. They must be able to dive to the bottom of the lake, return to the surface, and to repeat the process quickly. Fortunately, you can stop Buzz-Fizz from getting the food he wants to use. If you can find out: (1) what liquid is in the magic lake and (2) which food pieces are able to dive to the bottom of the lake again and again, the evil emperor can be stopped. After you find out what is in the lake and what kind of food dives best, an embargo can be placed on the food to keep Buzz-Fizz from getting it.

What to do: The spies that you have sent to Buzz-Fizz's country have just returned. They tell you that they have chemically analyzed the magic liquid, and it is a mixture of: three things: (1) water; (2) baking soda; (3) vinegar. Whew! Half of your task is done. The only other thing you now need to do is to find out which foods rise and fall in the magic liquid.

Guess which pieces you think will perform the best. After you've done this, fill a clear cup or drinking glass three-quarters full with water. Add a spoonful of baking soda to the glass; stir to dissolve it in the water. Slowly add 2 to 4 spoonfuls of white (distilled) vinegar to the glass. The solution will bubble right away! Take care that the liquid in the cup does not foam over the top! After you've finished adding the vinegar, the magic liquid in the glass will continue to bubble for several minutes. Place up to three different types of food pieces in the glass. Watch to see which food best repeatedly sinks and then returns to the surface. Which food works the best?

How it works: The food pieces that work the best will probably include (a) dry pieces of spaghetti, (b) unpopped popcorn, and (c) birdseed. A fourth food (raisins) work fine at first. But later they swell up with water and float uselessly on the surface, refusing to dive any more. Density (weight per unit volume relative to water) is very important in determining which food will work the best. The food must be slightly more dense than water, so that it will sink at first. After the food piece sinks, bubbles of carbon dioxide gas (from the reaction of vinegar with baking soda) will attach themselves to it. The food piece with attached bubbles is now slightly less dense than the liquid (mostly water), and rises to the surface. At the surface, the bubbles detach from the food piece, and the falling/rising cycle repeats.

More chemistry fun: Try using a clear carbonated drink (such as seltzer water) instead of baking soda + vinegar. Does the carbonated drink work as well? Also try dissolving an antacid tablet in water.

Grungy Green Pennies Thief-Catcher

You will need: pennies, paper towel, vinegar, dinner plate
The situation: For your birthday you received a brand new piggy bank. Since then, you've been saving all your pennies in it. But when you counted your pennies

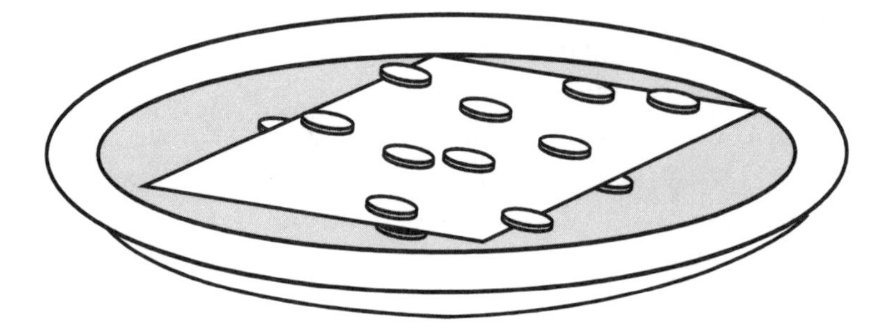

recently, it looked like some pennies were missing. Then you noticed that one of your brothers seemed to have more pennies than usual. When you asked him if he was taking your pennies, he said he wasn't. But pennies continued to disappear from your piggy bank. Finally, after several more weeks of vanishing pennies, you decided that you had had enough. You would catch the penny thief by marking your pennies in a way that they could be easily identified as yours.
What to do: Place several pennies on a paper towel that has been soaked in vinegar. Put some of the pennies on the edge of the towel, some entirely on top of it, and others underneath it. Let the pennies stand overnight. Take them from the towel and take note of the pattern of grungy green color on each one. Put these green pennies into your piggy bank, mixing them with the normal pennies. If any more pennies disappear, check for mis-

sing green pennies. If your brother has any green pennies, you'll know he's the penny thief!

How it works: Vinegar is a solution of acetic acid in water. Acetic acid reacts with the copper surface of the penny to give copper (cupric) acetate, which is a grungy green color.

More chemistry fun: Try contacting the pennies with other acids, such as lemon juice or sour salt (citric acid) in water. Do these acids discolor the pennies? Try leaving the pennies exposed to vinegar for shorter to longer periods of time (hours to days). How long must pennies be in contact with vinegar-soaked paper towels to give a green color?

Mysterious Multiplying Pennies

You will need: 100 to 200 U.S. pennies minted before 1982; 100 to 200 pennies minted after 1982; postal or kitchen scale, the weighing scale made earlier (page 17)

The situation: One of your friends offers you a deal. He will weigh out about a pound of his old pennies and exchange them for the same weight (about a pound) of your new pennies. He claims that his pound of old pennies probably contains lots of valuable coins. He has only requirement: you must repay him with pennies that are dated in the year 1983 or newer. He tells you that you are not allowed to count the pennies. It's either trade a pound of your new pennies for a pound of his old ones, or else he withdraws his deal and offers it to someone else. He says that the easiest way to assure a fair trade is to use the weighing scale made earlier in this book. He'll drop a pound of old pennies in one cup, and you drop a pound of new pennies in the other cup. When the two cups balance (the coat hanger is level), the weights of pennies in the cups are the same.

What to do: Suspicious about this deal, you obtain 200 to 400 pennies from the bank and divide the pennies into two piles: an old pile with pennies dated 1982 or before, and a new pile with pennies dated 1983 or after. You take an equal number of pennies from each pile. You borrow your family's postal or kitchen scale to weigh the piles. To your astonishment, you discover that the old pennies weigh *more* than the new pennies. So your friend was trying to trick you: he gives you fewer old pennies for a pound of your new pennies!

How it works: Pennies made before 1982 are made mostly of copper. Pennies made after 1982 are made mostly of zinc, with a very thin coating of copper. Because copper weighs more than zinc,* the older pennies weigh more than the newer pennies. You can use the weighing scale you made earlier in this book to see the difference.

*The density (weight per unit volume) of copper is 8.94 grams per cubic centimeter; the density of zinc is 7.14 grams per cubic centimeter.

Put ten old pennies on one side of the scale and ten new pennies on the other side. The side of the scale that contains the old pennies will go down, showing that old pennies are heavier than new pennies.

Weight comparison:

- 100 pennies minted before 1982 weigh 10.8 oz.
- The same number of pennies minted since 1982 weigh only 8.7 oz.

Number comparison:

- A pound of old pennies contains 152 pennies.
- A pound of new pennies contains a lot more: 185 pennies!

More chemistry fun: Put old pennies and new pennies in two separate glasses of clear vinegar. Check the glasses after several hours. Which type of penny (old or new) bubbles longer when in vinegar? Why?

Other Experiments

Wonderful Window Cleaner

You will need: empty window-cleaning squirt bottle, water, rubbing alcohol, household ammonia, food coloring, empty 1-gallon milk jug, small (3-oz) paper cups, dinner plate, vegetable shortening, lard or meat grease, paper towels

What to do: Don't pay a lot of money for store-bought window cleaner; make your own! Fill an empty 1-gallon milk jug half-full with water. Add a small amount of rubbing alcohol and ammonia in equal portions to the milk jug; start by adding just one small (3-oz) cup of each to the jug of water. Now test your cleaning solution. Coat a dinner plate with vegetable shortening, lard, or meat grease. Fill up an empty window-cleaning squirt bottle with your homemade cleaning solution. Squirt the solution onto the grease-covered dinner plate and wipe the plate with a paper towel. If the grease does not easily come off the dinner plate, add more rubbing alcohol and ammonia in equal small portions to the milk jug until the solution does remove grease. Write down how many cupfuls of each (rubbing alcohol and ammonia) it takes to produce a good grease-cleaner. Add food coloring to get the color of cleaning solution that you want (store-bought ones are usually blue, green, or pink). Congratulations! You've saved money by making your own window and glass cleaner!

How it works: Water dissolves dirt, but not grease and oils. Rubbing alcohol mixes completely with water and acts as a wetting agent; it helps water maintain contact with the surface to be cleaned. Ammonia does a good job of emulsifying (cutting through) grease. Although food coloring doesn't help the solution's cleaning power, it does make the solution look nice.

More chemistry fun: What happens if you use only one of the two active liquid ingredients (alcohol or ammonia)? Does your cleaner work as well as when both liquids are used in the cleaning solution? Which of the two liquids is a better grease-cleaner?

Foamy Salt and Soda

You will need: small paper cup, carbonated soda, salt, spoon, flour, powdered sugar, laundry detergent, and dishwashing liquid

What to do: Fill a small paper cup half-full with a carbonated soda. Watch the bubbles slowly rise to the surface. Count how many bubbles rise to the surface of the

soda every second. Now add a spoonful of salt to the cup. The soda will immediately foam! It may even foam enough to go over the top of the cup!

How it works: Salt loves water, readily taking up water and retaining it. This physical property is called *hygroscopicity.* To demonstrate this, simply leave some salt crystals in a dish for several days in a humid (very moist) room. The crystals will become wet, maybe even liquifying, if the air is humid enough.

Bubbles of carbon dioxide form in the carbonated soda according to the chemical equation:

H_2CO_3 (carbonic acid) → H_2O (water) + CO_2 (carbon dioxide).

Carbonic acid is unstable and wants to fall apart into water and carbon dioxide. Since salt loves water (H_2O), the equation is pulled to the right, speeding the evolution of the gas carbon dioxide (CO_2). Another reason that the liquid produces a foam is that carbon dioxide is less soluble in salt water than in plain water. When salt is added to soda, the salt dissolves, forcing out the dissolved carbon dioxide gas. One more reason is that salt changes the surface tension of water, making water molecules want to stick together more.

More chemistry fun: Try this experiment with other household chemicals, such as flour, powdered sugar, laundry detergent, and dishwashing liquid. Do any of them work as well as salt?

Relentless Rust

You will need: iron-containing objects (nails, steel wool, soap-covered steel wool cleaning pads, and paper clips), clear drinking glasses, water, salt, vinegar, dishwashing liquid

What to do: Place various iron-containing objects (nails, steel wool, soap-covered steel wool cleaning pads, and paper clips) into clear drinking glasses. Fill the glasses with water and allow the objects to stand for several days. Check them daily and note when rust first forms on them. At least one of these iron-containing objects (steel wool) has a coating that helps to protect against rust. Wash

one of the balls of steel wool with dishwashing liquid before putting the steel wool into water. It will rust faster. Try doing the same experiment in vinegar and water rather than water alone. Vinegar speeds up the rate of rusting.

How it works: Steel is made of iron with carbon in it. Rust is formed when iron combines with oxygen. The oxygen in air dissolves in water. In water, oxygen can react much faster with iron-containing materials than it can in air. Washing the protective coating from steel wool allows oxygen to react more readily with the iron. Vinegar speeds up the reaction of oxygen with iron in two ways: (1) by removing any coating on the iron material, exposing the iron metal itself; (2) by reacting with the iron itself, since vinegar is an acid, and iron reacts with acids, exposing more iron.

More chemistry fun: Try the experiment in vinegar solutions that are: (1) cold; (2) at room temperature; and (3) warm. How does heat affect the speed of rusting?

Food Color Combinations

You will need: bottles of artificial food coloring (blue, green, yellow, and red), water, clear drinking glasses

What to do: Fill two drinking glasses one-fourth full with water. Add 3 drops of blue artificial food coloring to one. Add 3 drops of yellow food coloring to another. Pour the contents of the glasses together. You will get a completely different color! Do this same experiment with all the rest of the other possible two food color combinations: (1) blue + green; (2) blue + red; (3) yellow + green; (4) yellow + red; (5) green + red. What colors did you get?

How it works: Red, blue, and yellow are called primary colors. All other colors can be made from mixing these three colors. Colors are what we see when light rays of a particular wavelength bounce off a substance into our eyes. Our eyes are very sensitive and are able to distinguish among a wide variety of light waves, whose wave-

lengths are very close together. The shortest wavelengths of visible light are found at the blue end of the spectrum. The longest are at the red end.

More chemistry fun: Mix colors together in different proportions to see how many new colors you can make. On the next page are some recipes for colors.

COLOR MIXING CHART

Number of drops needed				
Red	Yellow	Green	Blue	Mixed color
7	1			Strawberry
5			1	Rose
7	4	2		Light Brown
3	4	1		Toast
3	3			Rust
3	2	2		Chocolate Brown
3	2			Salmon
1	2			Orange
1	3			Peach
	12	1		Chartreuse
	3	1		Lime Green
	1	4		Pistachio Green
1	1			Olive Green
		1	4	Turquoise Blue
1		2		Blue Violet
3			1	Purple
1			2	Violet

For more science fun, pour one of the glasses of colored water into an ice cube tray and place the tray in the freezer section of your refrigerator. After they stand overnight, pop out the colored ice cubes onto a dinner plate. With a flashlight, shine a light on the cubes. Does the cube color change in brightness depending upon the direction you point the light? Which direction gives the most vivid color? Hold up an ice cube in front of your eye. Shine a flashlight through it toward your eye. How much brighter is the color in the ice cube with the flashlight pointed toward you than with the flashlight pointed away from you?

Food Coloring Streamers, Globules, and Diffusion

You will need: artificial food coloring, water, vegetable oil, clear drinking glasses, toothpick, soda water or uncolored carbonated beverage

What to do: Drop several drops of artificial food color into a glass of water, about 1 to 3 inches from the surface of the water. Notice how the color stays near the top of the glass. Continue watching. How long does it take for the color to reach the bottom of the glass? Add some vegetable oil to a second glass half-filled with water. Add a few drops of food coloring to the glass of oil and water. The drops of food coloring will act strangely. The

food coloring drops do not diffuse (spread out) right away as they had in the first glass of water, but rather form little globules (tiny balls) that float at the bottom of the oil layer, right above the water layer. Stir the globules with a toothpick until they break up. Notice that when they do break up, the lower (water) layer turns the color of the food coloring, but the top (oil) layer does not.

How it works: Artificial food coloring is a liquid which contains a colored chemical, water, and propylene glycol. In the first experiment (water only), the food coloring easily spread out, even though you did not stir the water. This is because water molecules constantly move around, rapidly colliding with each other. As the molecules do so, they carry food coloring molecules with them. In the second experiment (oil and water) the food coloring stays in the form of globules as long as it does because of the propylene glycol, which is soluble in both oil and water. Propylene glycol rolls into tiny balls, which are perfectly happy to stay suspended in the oil until they are punctured with the toothpick. Once they are broken up and stirred with the water layer, the globules of food coloring can easily dissolve in the water.

More chemistry fun: Do the first experiment in a glass of: (1) ice-cold water (with and without ice); (2) warm water; (3) hot water.

In which glass does food coloring diffuse the slowest? In which glass does it diffuse fastest? Do the same experiment with a saturated solution of salt water. (To saturate water with salt, add salt until the bottom of the glass of water is covered with salt. Stir. When, after stirring for at least a minute, no more salt dissolves, the salt water is saturated.) Does food coloring diffuse in salt water as quickly as it does in pure water? Add food coloring to a glass of soda water or uncolored carbonated beverage. Does the food coloring diffuse faster in soda water than in pure water? If so, why?

Water and Alcohol Color Dance

You will need: artificial food coloring, water, rubbing alcohol, dinner plate, tablespoon, soda straw

What to do: Mix three drops of an easily seen artificial food color (red, blue, or green, but not yellow) with three tablespoons of water on a dinner plate. Dip the end of a soda straw 1 inch into an open bottle of rubbing alcohol. Place your finger or thumb on the top end of the straw, lift the straw out of the bottle, and hold the straw over the top of the colored pool of water on the dinner plate. Lift up your thumb. The rubbing alcohol in the straw will drop into the water. The drops of alcohol will push a pocket into the center of the pool of water. As the water and alcohol mix together, the color will dance around. After a while (after the mixing has stopped) the color dance will stop as well.

How it works: The surface tensions of water and alcohol are different. Surface tension is the attraction that molecules in a liquid have for themselves. The skins of the two liquids (water and alcohol) are pulled at with differing amounts of force. Although water and alcohol are completely soluble in each other, it requires a few sec-

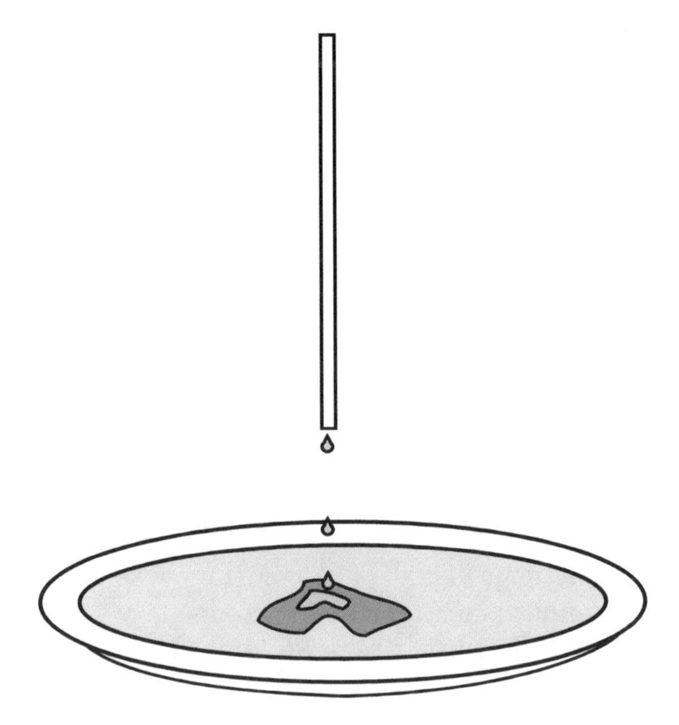

onds for them to completely mix together. While they are mixing, they push, pull, and tug at each other. The color dance of the alcohol in the water shows this mixing together.

More chemistry fun: The first experiment involved adding colorless rubbing alcohol to colored water. Do experiments that change: (a) what gets added to what (water or rubbing alcohol) and (b) which liquid is colored: (1) colorless water (in the straw) added to colored rubbing alcohol on the plate; (2) colored alcohol added to colorless water; (3) colored alcohol added to a different-colored water; (4) colored water added to colorless rubbing alcohol; (5) colored water added to a different-colored rubbing alcohol.

Try different ratios of rubbing alcohol to water. Does changing the ratio make any difference in the mixing behavior? For example, try 1 part alcohol to 10 parts water; then try 10 parts alcohol to 1 part water.

Disappearing Food Colors

You will need: beet, red cabbage, water, clear drinking glasses, liquid bleach, solild bleach, spoons, artificial food coloring, kitchen knife

What to do: Fill two clear drinking glasses with water. To the first glass add a tablespoon of liquid bleach. To the second glass do not add any bleach. Cut up a beet and soak the pieces in water for a couple of minutes. The red-colored liquid you get is beet extract. Add a spoonful of beet extract to each glass. The glass containing bleach will be lighter in color than the glass without bleach. Repeat the experiment with another colored food, such as red cabbage. Does red cabbage extract lose its color when mixed with bleach solution as much as beet extract does?

How it works: Liquid bleach contains a substance (an oxidizer called sodium hypochlorite) that destroys most colored chemicals, particularly those found in foods. When these colored chemicals are destroyed, their color disappears.

More chemistry fun: Do the experiment at a lower temperatures, in a glass of ice water. How much faster does beet or red cabbage color fade in a room-temperature glass of bleach solution than in a cold glass of bleach solution?

Use a solid bleach to decolorize food extracts. Solid bleach contains peroxygen bleaching agent, washing soda (sodium carbonate), and other ingredients. Solid bleach is milder than liquid bleach. Which bleach decolorizes beet and red cabbage extracts more quickly?

Mix an artificial food coloring in a glass of water with a spoonful of liquid or solid bleach. You probably will not see any fading with artificial food coloring, because the chemicals in artificial food coloring are much more stable than the chemicals in natural food extracts.

Kitchen Perfume-Making and Smelly Bubbles

You will need: good-smelling kitchen spices and extracts (vanilla, clove, cinnamon, mint, orange, lemon, banana), water, rubbing alcohol, small (3-oz) paper cups, pipe cleaner, dishwashing liquid

What to do: Pick out some good-smelling spices and extracts from your kitchen cupboard. Create a homemade perfume by mixing several drops of a spice or extract in a small (3-oz) cup with a small amount of water. Make additional perfumes in several different cups. To make even more perfumes, mix the contents of two or more of the cups. Try orange-mint, lemon-vanilla, etc. Use the combination that smells the best as a perfume or air freshener.

How it works: Certain plants, such as mint, have a very strong but pleasant smell or aroma. This aroma comes from molecules that are volatile enough to leave the surface of the plant and go into the air. The greater the number of molecules going into the air and the more responsive the human nose is to those particular molecules, the stronger the smell.

More chemistry fun: If you want a faster-evaporating per-

fume, use rubbing alcohol instead of water to dissolve kitchen spices and extracts. When you rub the perfume on you, on your friend, or perhaps even on your pet dog, how does it smell? But be careful if you apply your homemade perfume to a dog: their noses are much more sensitive than our human noses, and they may find the smell too strong to enjoy.

You can have a lot of fun sharing your perfume with your friends by blowing smelly bubbles at them (outdoors)! A particularly fragrant smelly bubble soap solution can be made by mixing 1 part dishwashing liquid, 10 parts water, and 20 drops of mint extract. Bend one end of a pipe cleaner into a loop. Holding up the straight end of the pipe cleaner, dip the loop end down into the bubble solution. To make lots of bubbles, pull the pipe cleaner loop out of the bubble solution and blow through it!

Paper Person in Submarine Glass

You will need: clear plastic drinking glass, piece of paper, flexible straw, bathtub, waterproof (duct) tape

What to do: Cut a small person out of a piece of paper. Rest this person on a wadded-up piece of paper. Wedge the person and then the wad of paper into the bottom of a clear plastic drinking glass. Turn the glass upside-down. Tape a flexible straw to the side of the glass at the bottom and middle of the straw with waterproof duct tape. The straw is your submarine's

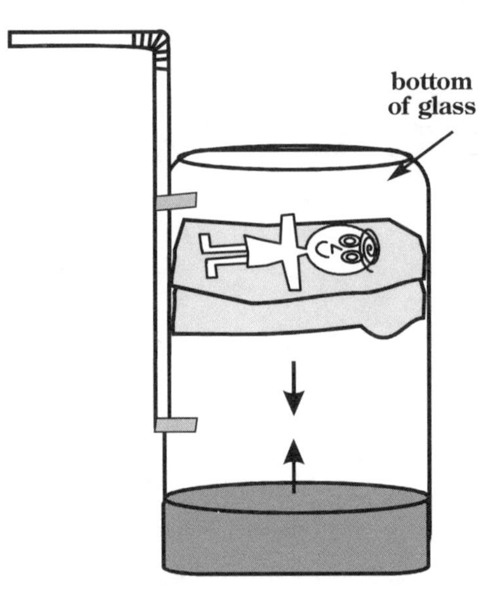

periscope. Take your person and submarine glass with you to the bathtub the next time you take a bath. Push the glass under water. The paper person will not get wet!

How it works: Only a small amount of water is forced into the glass when the glass is pushed below the surface of the water. Since the air in the glass cannot escape, it presses back against the water until the pressure of the air equals the pressure of the water. The wad of paper (and the person on top of it) wedged in the bottom of the glass is safely above the water level in the glass.

More science fun: Use taller or wider glasses. In which type of glass does water rise more? How deeply can you make your submarine go? To prevent the periscope filling up with water, place a piece of waterproof tape at the bottom of the straw.

Index